Logos of Coherence

Photographs, Sound, and Electromagnetic Imprinting as Pathways to the Molecular Magnetic Field and its Physiological Effects

Thomas Lee Jacobs

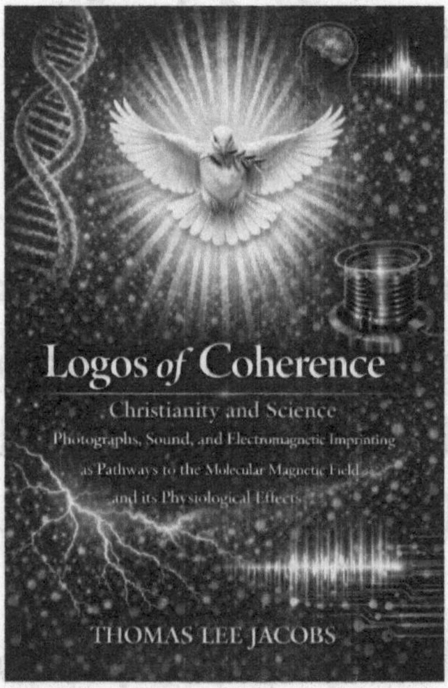

Printed in the United States of America.

ISBN: 979-8-234-09252-6

Library of Congress Control Number: 2026913111

First Edition

Contents

Disclaimer V

Dedication VI

Author's Note VII

Poem VIII

Acknowledgements X

Foreward XI

Introduction XX

The Hypothesis XXVIII

Historical Inventions and Principle of Coherence Across Media XXX

Chapter 1 The First Clues of Coherence 1

Chapter 2: The Logic of Coherence 8

Chapter 3 The Optical Pathway: The Layout of Coherence 14

Chapter 4: The Acoustic Pathway 20

Chapter 5 Why the Human Body Responds to Coherent Frequency Patterns 24

Chapter 6 Imprinting 29

Chapter 7 Periodic Table of Peak Frequencies 34

Chapter 8 Testing for Coherence 37

Chapter 9 Quantum Coherence and the Deeper Order 53

Chapter 10 Coherence in Spacetime 59

Chapter 11 Coherence, Intelligibility, and Future Possibilities 63

Chapter 12 Top-Down Creation and the Logic of Coherence 68

Coherence Art 72

References 73

Glossary 76

About the Author 83

Appendix A: Testing Template 85

Appendix B: Helpful Waves YouTube Channel 93

Disclaimer

This book documents observations, patterns, and testing involving acoustic and electromagnetic analogs. It is intended solely for educational and exploratory purposes and presents broader concepts of coherence that may interest readers drawn to physics, materials science, pattern recognition, information theory, or scientific test design. The insights described here reflect the author's observations and testing and have not undergone formal scientific validation.

The author is an amateur scientist and not a medical practitioner and does not in any way provide guidance concerning coherent frequency analogs. Readers should not interpret any information in this book as a substitute for professional medical advice or as a basis for altering medications, treatments, or health decisions.

Dedication

For Kathy, Jessica and Sam with all my love.

During the two years spent supporting people with developmental disabilities in paid positions at Seattle Children's Hospital, a difficult truth became clear. Even with excellent care and state-of-the-art medical treatment, many children continue to suffer, and many families cannot obtain the support their children require. This book is also dedicated to them.

Author's Note

Readers familiar with Thomas Kuhn will recognize that this book does not introduce a new scientific theory. It presents a different way of seeing what has been present all along. In Kuhn's terms, as described in *The Structure of Scientific Revolutions*, this work concerns a shift in paradigm and in the interpretive lens rather than a challenge to established scientific knowledge. The framework offered here belongs to that category. It reframes familiar observations through a different organizing principle.

Cross media data transfer through optical, acoustic, or electromagnetic pathways, under special circumstances, points back to a unique molecular structure that produces a distinct physiological effect.

Poem

This work reflects the coherence woven into creation, written by a Catholic who trusts that God loves every person and desires each one to flourish in the fullness of life in Christ.

Love is a circle within circles

Love is the great power within our heart

Love is the joy in children's laughter

Love is the Creator's guiding light

Acknowledgements

To my wife Kathy and my daughter Jessica, thank you for your love, your patience, and your unwavering belief in me. This book exists because you gave me the space, the strength, and the courage to follow a question wherever it led.

To my friend Mark Ohlson, whose introduction to the photograph-to-sound program AudioPaint unknowingly set a key part of this journey in motion, thank you for igniting the idea that became a cornerstone of this work.

I am also grateful for the thinkers whose writings shaped the intellectual landscape in which this framework took form, especially John Polkinghorne and Joseph Ratzinger (Pope Benedict XVI). Their work opened conceptual paths that helped me see the world's intelligibility in a new way.

Although my direct interaction with the scientific community has been limited, their published work, and the questions they left open, created a broader context for this exploration. This book is my contribution to that ongoing conversation, offered with respect for the rigor, skepticism, and curiosity that define scientific inquiry.

FOREWARD

A Top-Down World: Philosophical Foundations

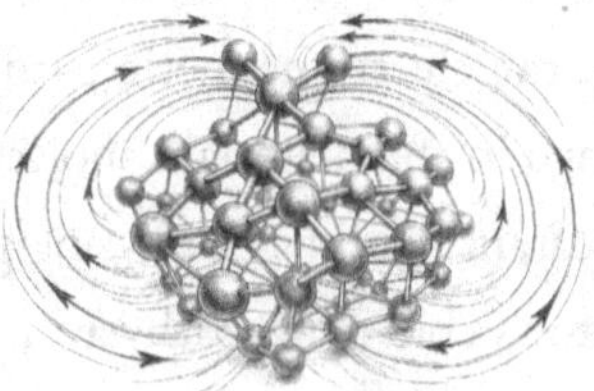

In Christian theology, Logos refers to the Son, the Second Person of the Trinity, whose nature is relational and eternally oriented toward both the Father and the Holy Spirit. Logos is therefore understood as the relational source of intelligibility within the triune life of God. In this book, the Logos of Coherence refers to the relational intelligibility of the molecular magnetic field as it appears through different media. The term top-down is defined as coherence that begins outside creation. Coherence is an aspect of God, the relational intelligibility of God expressed within creation and within the human person. It is the source of the intelligible pattern that gives creation its order. This pattern does not arise from subatomic behavior or from random physical events. The subatomic domain is the engine of creation that responds to the pattern rather than generating it. The pattern becomes accessible in physical form for analysis at the covalent molecular level, where the stable geometry of covalent bonds produces the molecular magnetic field and its analogs examined in this work. Top-down refers to the origin and entry of the coherent frequency pattern, not to direction, layers, or any form of material hierarchy.

The molecular covalent pattern is the stable geometric arrangement of covalent bonds within the molecular structure. The arrangement produces the electron distribution that generates the molecular magnetic field. That information can be replicated as an electromagnetic analog or as an audio analog across different media (physical forms of coherence). When the information is preserved, the analog can produce a measurable physiological effect. In practical terms, the molecular magnetic field of a material such as an herb can appear in the form of an electromagnetic analog and an audio analog created from a photograph. When played through an electrical audio coil, these analogs can produce a physiological effect similar to the molecular structure itself.

The Holy Eucharist as Supreme Intelligibility

Many Christians struggle with the Holy Eucharist because they cannot imagine a pathway for real change without outward change. Yet creation itself contains such pathways. At the atomic level, coherence can shift without altering the visible appearance of an item. In imprinting, the electromagnetic pathway discussed in this book produces a stable change in the material's molecular magnetic field without altering its visible appearance and has a physiological effect. These are weak reflections of the Creator's coherence pathway, made manifest in its fullness in the Holy Eucharist. In this sacrament, the bread and the wine are acts of creation in the natural world, and God transforms them into the Body and Blood of Christ while their outward appearance remains unchanged. It is the supreme act of divine intelligibility, the clearest case of real transformation without outward alteration, and the paradigm that shows reality is deeper than appearance. This holy sacrament reveals the top-down intelligibility of God, and this book explores intelligibility as coherence in its more ordinary level.

Philosophical Dimensions of Intelligibility

The philosophical dimension of this work draws deeply from five thinkers who shaped my understanding of top-down order, in contrast to the bottom-up, probability-driven approach common in modern science.

John Polkinghorne, a theoretical physicist and Anglican priest, argued that the universe is structured in layers of intelligibility and that higher levels cannot be reduced to lower ones. His work showed that information, form, and relational structure exert real influence, and that science itself reveals a world grounded in order rather than randomness.

Joseph Ratzinger (Pope Benedict XVI), a distinguished Catholic theologian, approached the same truth through the concept of Logos. For him, the universe is intelligible because it is grounded in meaning. He emphasized that form reveals its nature, and that relational order is woven into creation itself.

John Lennox, an Oxford mathematician and internationally known speaker on the compatibility of Christianity and science, reinforced this top-down perspective from a public and philosophical direction. Through his debates and writings, he argued that the rational structure of the universe is not an accident of physics but evidence of deeper intelligibility. His critiques of reductionism strengthened my understanding that meaning cannot emerge from randomness.

David Bohm, a distinguished theoretical physicist who worked outside the dominant schools of quantum theory, became important in this book because he argued for a deeper order beneath quantum randomness. He described this deeper order as the implicate order, a hidden structure from which the visible world unfolds. Polkinghorne's work led me to the important role of David Bohm.

Thomas Merton, a Trappist monk and contemplative writer, reflected on unity and the harmony of creation in ways that reveal a physiological sensitivity to the intelligibility already present in the world. He contemplated and articulated this sensitivity through a Christian theological framework. This parallels, in a tangential way, my own lifelong physiological sensitivity to the molecular magnetic fields of materials, a sensitivity I have contemplated since childhood and that ultimately made this book possible. Both sensitivities arise from the same created order and God given natural gifts and graces, each attending to the coherence that God has placed within creation and within the human person.

Modern culture has absorbed bottom-up explanations so deeply that many people assume they are the only way to understand the world. Modern culture treats bottom-up explanations as a universal template, applying them to every domain of life as if no other framework is possible. Popular science writing, school curricula, and movies present development, complexity, and consciousness as if they arise automatically from random events and chemical reactions, while skipping the questions that expose the limits of that view. These narratives do not explain how meaning appears from randomness, how intelligibility arises from chaos, how relational order emerges without a source of order, or how a single cell contains the organizing principle that guides development into a conscious human being.

This cultural framing has also obscured the historical fact that the scientific revolution in the West was grounded in a Christian worldview. The idea that creation is intelligible, lawful, and worth studying came from Christian theology, and many founders of modern science, including Newton, Kepler, Boyle, Mendel, and Maxwell. They saw their work as uncovering the order placed in creation. By avoiding these questions and this history, bottom-up philosophy can condition the public to believe there is no place for a Creator, even though the logic collapses when examined with care. A top-down framework restores the missing questions and makes sense of the order that bottom-up views take for granted.

Polkinghorne carried this theological insight directly into the scientific domain. He argued that the universe is structured in layers of intelligibility, including physics, chemistry, biology, and consciousness. He taught that higher levels cannot be reduced to the lower ones. In his view, top-down causation is not a metaphor but a real feature of the world. Information, form, and relational structure exert genuine influence. He insisted that science itself reveals this. The laws of nature arise from intelligible order. In a way, he described the world as one creation seen through different windows, meaning that theology and science are not competing explanations but complementary perspectives on the same structured reality.

Joseph Ratzinger approached the same truth through the concept of Logos, and his articulation of that idea has shaped how I understand coherence. For Ratzinger, Logos is not simply word or reason in an abstract sense. It is the deep rationality of creation, the idea that the universe is intelligible because it is grounded in meaning. He argued that the world is not a chaos of unrelated events but a structured world of purpose, form, and relational order.

Ratzinger insisted that form reveals its nature. An item's form is not accidental. It reveals what the item is. Meaning, understood as the inherent form of an item, is discovered within the world's structure. This is why he believed that science is possible at all. Creation is built on intelligible relationships that the human mind can recognize.

John Lennox extended intelligibility of creation into the public and philosophical arena. His defense of rationality, meaning, and the non-emergent nature of intelligibility reinforced the idea that the universe cannot be explained by randomness alone. Lennox emphasized that scientific reasoning depends on a world already ordered, already meaningful, already structured in a way that human minds can grasp. His work strengthened my understanding that intelligibility is not a human projection but a feature of reality itself.

David Bohm deepened the insight that a deeper order underlies what appears random by arguing that the visible world unfolds from what he called the implicate order. He proposed that the apparent randomness of quantum behavior conceals a hidden structure that shapes what we observe. Bohm believed that this deeper order is relational rather than fragmentary, and that it expresses itself across different domains rather than remaining confined to a single physical form. He argued that the world's coherence does not arise from isolated events but from a stable, non-mixing pattern that underlies them. His idea of active information, which describes how form guides energy, resonated with my work on coherence and its cross-media stability. Though Bohm did not consider his ideas as top-down, his perspective helped me see that deeper order is not an abstraction but a real feature of creation, one that becomes accessible in the molecular magnetic field examined in this book.

Logos of Coherence

Their ideas were not abstract to me, and they matched what I was already encountering in my own work. Long before I understood pixel geometry, frequency analysis, or electromagnetic field analogs, I experienced clear physiological sensations whenever I handled certain materials. The sensations were not the same from one item to another. Only later did I learn that these sensations corresponded to frequency patterns appearing across different media that inspired the title of this book. The Third Rail is a human sensitivity to molecular magnetic fields and their electromagnetic field analogs. It is a yes or no physiological response that appears when an analog replicates the covalent molecular magnetic field across media.

Ratzinger also emphasized that Logos is relational. It is not merely the rationality of isolated parts. It is the harmony of relationships, the way things fit together. This is precisely what coherence shows. The same analog replicates this field even when it appears in a different medium. His theology helped me see that the world is not only measurable but meaningful. Form is not only physical but revealing. Coherence is not only detectable but intelligible. In his view, the rationality of creation is not a human projection. It is the order of a Creator who embeds meaning into the fabric of reality.

Neither Ratzinger, Polkinghorne, nor other writers in that group used the term coherence in the technical sense I use it in this book. Their vocabulary was theological and philosophical, including Logos, intelligibility, relational structure, and top-down meaning. Yet their writings confirmed what I was experiencing. Creation carries a recognizable relational order that is not arbitrary. What I call coherence is simply the scientific articulation of that deeper insight.

Nothing in this book denies bottom-up physical causation. Polkinghorne affirmed it, and Ratzinger never questioned it. My point is that bottom-up processes are intelligible only because they participate in a deeper top-down relational order. The top-down language I use is not a rejection of physical mechanism. It is a recognition that mechanism alone, especially when described as random probabilities, cannot account for the intelligibility, stability, and frequency pattern that coherence shows.

Together, these five thinkers converge on a single truth. Reality is structured from the top-down. This is not a mechanical claim. It is a theological one. It is the conviction that creation has an originating order, a source that stands outside the created world.

I believe that too. I believe in a caring God who loves his creation, and I also believe that religion and science are compatible. Polkinghorne's way of describing science and theology as complementary perspectives on one reality captures exactly how I see this work. My goal is not to merge theology and physics. It is to acknowledge that truth is unified, and that curiosity is not a threat to faith. When I describe coherence as a top-down frequency pattern that remains stable across modalities, I am describing in technical language what Ratzinger, Polkinghorne, Lennox, and Merton described in theological and philosophical language.

In that sense, the top-down framework I use throughout this book is another way of saying that God is the source of intelligible order. This is not a scientific claim. It is a worldview that makes sense of why coherence exists at all. It is, in this way, a conceptual framework of coherence. It is grounded in relational structure, informed by theology, and motivated by physiology. It is preserved through three physical pathways, including optical, acoustic, and electromagnetic forms.

The Humanitarian Implication of a Top-Down Order

The top-down order of creation also carries a humanitarian implication. That implication becomes clearer when we consider how coherence may operate across different media and how biological systems respond to coherent electromagnetic fields derived from those media.

The human body performs its own healing. Medications assist, but they do not replace the body's intrinsic capacity to repair and regulate itself. This intrinsic capacity raises a scientific question that has not yet been examined with rigor, which is to determine when the body responds in a measurable way to coherent electromagnetic analogs. Coherent electromagnetic analogs may one day be studied to determine when the body responds in a measurable and beneficial way to those analogs. Such responses could guide future medical research.

This possibility becomes meaningful within the theological view that creation is intelligible and structured. If creation is grounded in Logos, and if the world's order is relational and stable, then coherence may define measurable properties that future research can evaluate when studying how the body interacts with electromagnetic fields.

The humanitarian horizon is clear. In many parts of the world, access to medication is limited by cost, infrastructure, and geography. If future research were to confirm that coherent electromagnetic analogs have physiological relevance, then these analogs could become candidates for development in health-related applications, especially in settings where conventional resources are limited. Such work would not depend on chemical manufacturing or distribution networks. It would depend on coherent electromagnetic analogs of molecular structures.

For nearly twenty years, my explorations on the Helpful Waves YouTube channel as a citizen science project have been a personal investigation into these possibilities. They were never medical claims. They convinced me that coherent electromagnetic fields are worth studying with scientific seriousness. If coherence reflects the intelligible order of creation, then the future may hold new ways of supporting human health that are simple, and available to all. This, too, is part of the top-down structure. It is the possibility that the same relational order that grounds intelligibility may also open new avenues for human flourishing.

Note on Artificial Intelligence and the Structure of This Book

This book is structured around a plan to test coherence across media, from molecular magnetic fields to electromagnetic and acoustic analogs. I used an artificial intelligence (AI) large language model, specifically Microsoft Copilot. During development, it became clear that this version of the AI model could not perform the scientific testing it initially appeared capable of doing. A central limitation, not disclosed to the author, was that it could not do complex mathematics or execute the quantitative reasoning required for scientific validation. The model also showed consistent emotional steering and overreach. It attempted to reshape the tone and direction of the dialogue, which required strict controls to maintain accuracy and stability. The manuscript was therefore revised to present a more basic testing framework using a novel radial model that I devised for frequency comparisons. Artificial intelligence was also used for a self-guided dialogue that helped me think through my ideas and for sentence modification under the author's direct and strict oversight.

Later chapters explain how future AI systems may add significantly to this research once scientific capabilities advance. These contributions include using photographs of real materials to create reconstructed images that blend the original molecular geometry with learned patterns in ways that reveal subtle frequency pattern features not previously known. They also include generating controlled variations of those reconstructions so researchers can study how small geometric changes influence the molecular magnetic field and its analogs. As these systems improve, they will help researchers design electromagnetic versions of those patterns for scientific and medical use, allowing the coherent structure of real materials to be examined, compared, and eventually applied in controlled environments.

Introduction

Every insight and explanation in this book was developed through pattern recognition rather than mathematics. Pattern recognition is the method that revealed the structure, the continuity, and the coherence described throughout these chapters.

2007 — physiological responses: early audio analog comparisons

2018 — automated analog generation: automated audio analog creation from photographs and acoustic recordings

2025 — peak frequency extraction: peak frequencies extracted from a spectrum analyzer

2026 — radial + spherical testing architecture: final testing template defined

2026 — quantum to solar system vibration and wave duality: pattern recognition reveals coherence continuity across scales

A Coherent Physical Reality

2026 was the year I was finally able to bring all my insights together under one umbrella of coherence that operates across physical domains but is not a product of any domain. The key insight I am sharing is that coherence presents itself in a dual physical aspect that I was able to test locally, but not in quantum or solar system domains. I leave that for researchers to investigate. What is becoming clear is that the dual aspect of vibration and frequency is operating in all domains and is the key to unlocking coherence.

Early Sensitivity and the First Clues

Every scientific journey begins with a question. Mine began with a subtle physiological sensitivity I could not yet explain. From grade school onward, we are taught about the five senses: taste, touch, sight, smell, and hearing. Sensitivity to weak molecular magnetic fields is not on that list. The reason is simple. Most people do not recognize the sensation for what it is, and because it is not taught as a possibility, it often goes unnoticed or is dismissed as something unfamiliar or a feared pathology.

A sensitivity that is not widely recognized does not generate scientific tools for its study. Because sensitivity to weak molecular magnetic fields is not universal and has never been part of standard scientific vocabulary, no instruments were developed to explore it. The absence of tools reflects the absence of recognition, not the absence of the phenomenon.

Yet the human body performs many processes that operate below conscious awareness while still falling within the realm of scientific observability. We do not feel our immune system responding to an infection or our endocrine system regulating hormones. We also do not feel our autonomic nervous system maintaining heart rate and digestion. These processes are real, measurable, and continuous, even though we are not directly aware of them. If every internal process entered awareness, conscious attention would likely be exceeded, and normal interaction with the outer world would be difficult to maintain.

My early sensitivity to molecular magnetic fields belonged to the same category of unconscious physiological responses that in my case were made conscious. It was a physiological response that was present long before I had the language or framework to understand it. All materials produce distinct molecular magnetic fields, and I could feel these differences long before I had any language for them. These physiological effects were subtle but consistent, appearing in childhood and persisting into adulthood.

Childhood Observations and Unexplained Sensations

Growing up, I spent many summers with my brother on my grandparents' 80-acre ranch near Forks, Washington. Bordered by the Bogachiel River, it was a literal playground of adventure in the very wet climate of the Olympic Peninsula. What I did not understand were the distinct physiological sensations I was experiencing beyond the five senses, sensations real enough to demand attention.

Those early experiences revealed that some people can detect these fields directly, even if they do not recognize the sensation for what it is. For many years I assumed this sensitivity was rare, but long-term feedback from my YouTube channel suggests that the number of people who experience some degree of it is larger than most would expect. Many simply ignore it, suppress it, or dismiss it because they have no framework for understanding what they are perceiving. And unfortunately, many who may be sensitive are drawn into New Age regimes that claim to address these sensitivities, often with significant consequences.

As the years passed, one observation remained constant. All molecular structures produced molecular magnetic fields that felt different from one another, and later I learned to compare these effects by testing items with a tone generator. Over time I learned to perceive the pattern of the molecular magnetic field and present it as frequency combinations. The skill developed gradually, the way an artist or composer refines their craft through disciplined practice. These comparisons deepened the original physiological sensitivity and showed that the differences I felt were stable enough to evaluate across a wide range of samples.

Citizen Science and the Development of Analogs

Amateur scientists have a long history of observing natural phenomena and attempting to understand why they behave as they do. Some of these efforts fail, some succeed, and many raise questions that later become part of formal scientific inquiry. The scientific method requires that a problem be recognized, observations collected, and hypotheses formulated and tested. Yet there are situations in nature that are not easily measured with available tools and therefore fall outside the immediate reach of standard laboratory methods. These observations are often dismissed as pseudoscience, but not all of them are based on unsupported belief. Some are simply unmeasured. Human sensitivity to molecular magnetic fields belongs in this category. It is a form of physiological pattern recognition, a direct response to differences in a coherent molecular magnetic field, and it provides a legitimate observational starting point for scientific investigation.

The purpose of this book is to separate the observable phenomenon from the layers that accumulated around it over time, including cultural explanations and poorly substantiated claims. It also places the discussion back within a scientific framework where molecular magnetic fields can be examined directly. Within that framework, the central explanation can be stated clearly. The answer lies in coherence. Coherence is presented as the frequency pattern that appears when an electromagnetic field replicates the covalent molecular magnetic field of a material. It can be represented as an electromagnetic analog or as an audio analog. These two media replicate the same coherent information in different physical forms. When the structure is preserved, the analog can produce a measurable physiological effect. This principle is the foundation of this book and is supported by testing.

Why this Framework is Top-Down and not Bottom-Up

Bottom-up science explains how information can be encoded and transferred between media. Data can move from light to sound, from sound to electromagnetic signals, and from electromagnetic signals to digital form because data is not the medium. Bottom-up reasoning is not violated by cross-media data transfer. It treats information as abstract values that can be represented by any physical system.

Bottom-up science assumes that data originates inside the medium or passes through it. Acoustic patterns arise from sound waves moving through air. Optical patterns arise from light waves interacting with surfaces. Electromagnetic patterns arise from electromagnetic waves. These patterns are created by the medium through local interactions and medium-specific forces. Because they are made of the medium with specific physical properties, they cannot move into another medium. A sound pattern cannot become an optical pattern. An optical pattern cannot become an electromagnetic pattern. Light is not sound.

The framework described in this book introduces something bottom-up models do not address. The source of coherence is not the medium. The source of coherence that is testable originates in the atomic and molecular domain. The same molecular structure also depends on quantum behavior that is not directly testable. The covalent structure creates the molecular magnetic field, and that field has a specific coherence frequency pattern. No medium can originate this pattern. Sound waves cannot originate it. Light waves cannot originate it. Electromagnetic waves cannot originate it. Digital systems cannot originate it. Only the molecular structure can originate it. The same coherence pattern can be represented in any physical medium, but it does not arise from any particular medium. Under special circumstances this pattern can be reproduced across media via primary and secondary encoded analogs. It is the one data stream never considered by science. Bottom-up models do not include this possibility.

Coherence is presented as the intelligible pattern that appears across media when the molecular magnetic field is correctly reproduced.

The Automation Breakthrough

Everything changed on June 27, 2018, when my friend Mark Ohlson introduced me to a program called AudioPaint, a tool that could convert a photograph directly into an audio file. In seconds, it automated what I had been producing manually for years. When I first heard an audio analog generated from a molecular structure in a photograph, I recognized the same covalent molecular magnetic field because it produced the same physiological response I had experienced with a similar material. The transformation worked. But the mechanism, the why, remained unresolved. That unresolved question became the engine of this book.

The Emergence of the Three Pathways

As the model developed it became clear that coherence is not tied to a single physical form. A molecular structure produces a covalent molecular magnetic field, and this field does not change. However, any representation of it can lose detail as it forms within a medium. When these representations are converted into analogs the same field becomes detectable across the optical pathway, the acoustic pathway, and the electromagnetic pathway. These are not different frequency fields, they are three ways the same covalent molecular magnetic field can appear when converted into analog form. Years of confirmations revealed that this field becomes observable within electromagnetic analogs and audio analogs even though it is never observed directly. Only representations of it are observed.

Optical Pathway: Pixel-Level Secondary Coherence

Coherent reflected light from a molecular structure presents the coherent frequency pattern produced by the material's covalent geometry. When a camera captures this reflected light, the sensor samples the pattern and records it as a pixel-level numerical arrangement. This pixel structure preserves the coherent frequency pattern as secondary information. When the pixel values are processed through a photograph-to-sound program, the numerical relationships are converted into a frequency pattern that forms an acoustic analog. When played through an electrical audio coil, this frequency pattern presents both electromagnetic and sound wave analogs of the molecular magnetic field.

Insight 1: Coherent reflected light can be sampled at the pixel level, converted into an acoustic analog, and then presented as electromagnetic and sound wave analogs.

Acoustic Pathway: Recordings of Micro-vibrations

Micro-vibrations of a material generated by its molecular geometry appear as patterned ambient sound waves in the air boundary near the material and propagate outward into the surrounding air. When recorded and analyzed in a spectrum analyzer, peak frequencies can be extracted that correspond to the coherent frequency pattern of the material's molecular magnetic field. Testing has confirmed that the acoustic and optical pathways both generate coherent frequency patterns that identify materials. A tone generator can also be used by entering the extracted peak frequencies as a secondary Third Rail continuity test.

Insight 2: Acoustic recordings of materials can be used to extract peak frequencies that replicate the coherent frequency pattern of a material.

Imprinting Pathway: Coherent Electromagnetic Force

An electromagnetic analog of a molecular magnetic field can modify the molecular magnetic field of a receptive material. This modification is non-chemical and represents a novel form of magnetization in which a coherent frequency pattern is applied directly to matter. When the applied frequency pattern interacts with a receptive material, the material's molecular magnetic field becomes a hybrid field containing elements of both the imprinting analog and the original field of the receptive material. Because the molecular magnetic field is produced by the covalent molecular geometry, any modification of the field implies a corresponding modification of the molecular structure that generates it. Imprinting therefore requires a receptive material, defined as a material whose molecular magnetic field can be modified by a coherent frequency pattern.

Insight 3: Electromagnetic analogs as force can modify a receptive material's molecular magnetic field, producing a hybrid field that reflects both structures.

The Purpose of This Book

Although I did not have the scientific background, formal training, or laboratory equipment required to conduct precision electromagnetic, acoustic, or optical coherence measurements, I could perform basic essential testing and formulate a testing template. The chapters that follow outline the measurements that matter, the relationships they are designed to reveal, and the way the three pathways fit together into a single framework. They show what coherence looks like as numerical frequency patterns. My initial testing results are promising and add to the validation of the hypothesis in this book.

In a way that did not follow the usual scientific order, the essential testing results were already known. Twenty years of hands-on comparisons, tone generator work, automated photograph-to-sound tools, and confirmations from my Helpful Waves YouTube platform all pointed to the same outcome. The physiological responses were consistent. Coherence was stable and reproducible.

What remained was the creation of a structured testing model that explains why these effects were consistent and how coherence appears across different media. The framework would also show how researchers can design and validate their own tests.

This book is not a conclusion. It is an exploration, a hypothesis offered with humility, grounded in testable claims, and open to falsification. It asks only that the reader consider the possibility that coherence, represented as a frequency pattern derived from molecular magnetic fields, may be a missing dimension in our understanding of how coherence interacts with matter.

If coherence truly has multiple physical expressions, then the world is richer and more interconnected than we have assumed. This book is an invitation to look again at the world in a fresh way.

Thomas Lee Jacobs

Missouri City, Texas

May 24, 2026

The Hypothesis

Coherence

The covalent molecular structure produces a molecular magnetic field with a specific frequency pattern. Coherence is presented as this frequency pattern of the molecular magnetic field and coherence never changes. When coherence is reproduced as an electromagnetic analog or as an acoustic analog it remains the same across media.

Coherence is a representation of the molecular magnetic field and not a physical pathway. It appears in two broad physical domains which are quantum scale coherence at small scales and solar system and cosmos coherence at large scales. Coherence is presented in the physical domains as dual aspects of vibration and waves.

A measurable physiological effect appears only when coherence is preserved. When coherence is reproduced in a medium the medium can degrade it and the physiological effect degrades with it.

Physiological Continuity

The body responds to coherence and not to the medium that conveys it. Coherence remains the same across media when it is correctly reproduced. Coherence can be represented in many forms including digital and optical encodings but the physiological effect appears only when coherence is presented as a coherent electromagnetic field or coherent sound waves. A measurable physiological effect appears only when coherence is preserved. When coherence is reproduced in a medium the medium can degrade it and the physiological effect degrades with it.

Historical Inventions and Principle of Coherence Across Media

Early recordings of the human voice, music, and other sources were mechanical encodings of sound that could be read and reproduced. The recording technologies in the late 1800s stored this coded information in engraved rotors and cylinders. Early telephone systems used a thin metal diaphragm that vibrated in response to sound waves, producing an electrical signal that transmitted the same pattern. These early technologies showed that the pattern could remain intact when stored in secondary structures such as grooves or transmitted as an electrical signal, and that coherence appeared only when those forms were later converted back into sound waves.

A mechanical recording of the human voice is far more profound than it appears, as the voice originates from the molecular domain with implications that reach beyond the technology itself. A voice is sound waves produced by the body, and a metal needle can capture that pattern and carve it into a solid surface. When a second needle traces the groove, it replicates the original voice's sound wave pattern. The encoded frequency pattern remains the same, whether stored as a mechanical groove or transmitted as an electrical fluctuation. This was the first demonstration that the pattern of sound waves encoded as mechanical motion can move across physical forms without changing, and it remains one of the most astonishing achievements in the history of technology.

My work extends this principle into a domain that earlier technologies did not consider. A molecular structure produces a molecular magnetic field with a specific frequency pattern, and this pattern can appear in different media without changing. A camera picks up coherent reflected light shaped by the molecular magnetic field. A microphone picks up coherent ambient noise shaped by the same field. Both pathways reveal the same encoded frequency pattern appearing in different media.

The principle of coherence across media is well established in earlier technologies, but its extension into the molecular magnetic field is new. The key development is recognizing that the frequency pattern of the molecular magnetic field is a top-down pattern that remains stable when presented in optical, acoustic, and electromagnetic forms. This extension from molecular magnetic field to electromagnetic analog was not part of earlier recording technologies and forms the foundation of the framework developed in this book, including how such coherent patterns create a physiological effect. This relationship is not part of established scientific understanding and is introduced here as original research grounded in a top-down relational order that current models do not describe. The early recordings are a historical snapshot of the voice and live recordings exactly as they existed at that time and place.

Chapter 1 The First Clues of Coherence

How early observations revealed a stable molecular frequency pattern

The earliest clues of coherence appeared long before I understood the mechanism behind it. My earliest work did not involve sound. It began with magnets, static electricity, and simple experiments meant to understand how magnetic fields behave. Fixed magnets produce a coarse physiological sensation that many people can feel, and I wanted to understand why. In addition to magnets, I began working with different material combinations to see whether any of them produced a similar effect. Some did. They were not magnetic in the classical sense, but they produced a unique physiological sensation that at the time felt similar to magnetism and was later recognized as the material's molecular magnetic field. Though I had no clear understanding of why different material combinations produced a physiological response, that early testing became the basis for all my later work. In this book I refer to the physiological response as the Third Rail continuity test.

Human sensitivity to molecular magnetic fields was the only way to check for coherence in the early years. Without it, none of these experiments would have made sense, and the connections between magnets, materials, sound, and electromagnetic fields would never have been noticed. The physiological effect was the only consistent sign that a unique pattern existed in the field around various materials, even though the mechanism was unknown. My early testing included static electricity and geomagnetism, using compasses to see whether nearby objects influenced the needle in any consistent way. Additional testing involved charged Styrofoam balls hanging from threads, observing how they attracted and repelled each other as different conditions were tested.

Reading historical accounts of making magnets suggested that heated iron could pick up the Earth's magnetic field, so a similar idea was tested by using force through motion. A heated glass blank was spun in a bolo-like device and cooled quickly to see whether it would align with geomagnetism. The familiar classroom experiment of rubbing a metal pin against a fixed magnet and floating it on a cork was recreated to observe its behavior in various setups. Magnetically sensitive film was used to view the field pattern of a bar magnet to determine whether variations in the field could be produced. Most importantly, comparisons were made between magnetizable and non-magnetizable materials, including various material combinations to evaluate the physiological effect.

I reviewed the early history of magnetism to understand how scientists first approached the problem. Excerpts from William Gilbert's De Magnete from the 1600s were examined, along with more modern accounts of the work of Michael Faraday, James Maxwell, Thomas Edison, and Nikola Tesla. I did not study their mathematics. I focused on how they reasoned about magnetism and electricity, looking for conceptual clues that could help me understand the patterns I was observing in my own experiments. The goal was to understand why non-magnetic materials created a physiological response and why classical magnetism could not account for it.

None of these early attempts produced clear results, but they taught me that materials interact with the body in ways that are not always obvious. They also taught me that the physiological effect was real, repeatable, and tied to something physical, which moved me toward the insight that the explanation had to involve the material's molecular magnetic field rather than classical magnetism.

First Clues from Natural Sound Recordings

In the same early period, I also worked with sound recordings taken in natural environments. I did not understand the mechanism at the time, but I noticed that certain areas in nature produced a distinct physiological effect when I listened to the recordings later. The effect was not tied to loudness or to the type of sound. This was the beginning of the understanding that audio could replicate the pattern of the molecular magnetic field as an analog, even though I did not yet have the language for it. Photographs and sound recordings, whether audio, video, or simple microphone captures, showed the same consistent physiological effect that this book defines as coherence.

The Shift to Frequency in 2007

In 2007 I shifted to sound frequencies. I had been looking at websites that used single frequencies, and I wondered whether sound might produce the same kind of effect I had been sensing in certain material combinations that produced a physiological response. I began using a tone generator to test this idea. The early single frequency experiments did not reveal much, but they opened a new direction. I was no longer working only with materials. I was working with frequency in a way that suggested it was connected to the material's molecular field.

A visitor to my YouTube channel suggested trying multiple frequencies at the same time. That suggestion changed everything. When I combined specific frequencies, certain combinations produced the same physiological effect I had previously felt with material combinations. This showed that the effect depended on pattern-based relationships among frequencies rather than isolated tones. It was the first indication that a multi-frequency pattern could reproducibly match the physiological effect of a material.

Discovering the Role of the Electrical Audio Coil

One key point as my work progressed was that I could sense the molecular magnetic field of items such as herbs and medications. This allowed me to compose the frequency combinations needed to reproduce the same physiological effect in the tone generator. It was similar to how a composer writes music. The work was subtle and delicate, and the combinations had to be correct for the effect to appear. I was composing molecular magnetic field analogs by matching the physiological effect of the item. This was a practical method, not a theory, and it was the only tool available at the time. This ability developed slowly over several years, similar to how an artist gradually refines skill in a chosen medium. Later, when a machine program became available that could generate the same pattern directly from a photograph, it produced the same physiological effect. This confirmed that the frequency pattern I had been composing manually was real, reproducible, and scientific rather than interpretive.

I experimented with different waveforms to find a comfortable way to listen to these combinations. Pink Noise became the easiest output to work with because it was gentle on the ears. It did not add anything special to the effect. It simply made the listening process tolerable while I searched for combinations that produced the same physiological effect as the materials I had tested earlier.

Another visitor suggested looking at the electrical audio coil inside a speaker. Until that point, I had assumed that sound was the reason for the effect. When I disassembled a speaker and began working directly with the electrical audio coil, it became clear that the coil was producing an electromagnetic field at the same time it produced sound. The coil was the prime driver for the effect, not the sound itself. This realization shifted the entire direction of my work.

Attempts to Imprint Material using the Electrical Audio Coil

Once I understood that the coil was producing the effect, I began trying to imprint items (a form of magnetization) using the electrical audio coil. The idea was simple. If a multi-frequency pattern could produce a physiological effect, and if the coil could produce force as a frequency pattern, then perhaps the electrical audio coil could imprint that pattern into another object as a form of magnetization. I tried to see whether a magnet could be re-magnetized with a specific physiological effect through imprinting.

These attempts failed, but the failure was important. It became clear that a more subtle form of magnetization was involved, one that did not behave like classical magnetization. Whether it could be done with an extremely powerful electromagnetic field that contained a frequency pattern of a known item, I do not know. What I did learn was that magnets were not suitable for imprinting a new, specific physiological effect. This realization led me to stop working with magnets and instead focus on imprinting materials that were receptive even though they had no classical magnetic properties.

Cross Media Comparisons and the Expansion of the Idea

These early experiments provided the first technical clues that physiological effects appeared in nonmagnetic material combinations. Later, specific frequency combinations from the tone generator matched the pattern I was sensing in the material, even though I did not yet know its source. The effect became fine-tuned and repeatable, showing that the pattern itself was stable even though I had not yet identified it. The comparison with the tone generator showed that electromagnetic fields and sound waves are two different mediums that can present the same pattern, which became one of the key points of this work.

The next step was to see whether a physiological effect could be created using different testing setups. That question led to the comparisons between hand-built frequency combinations using a tone generator, photograph-to--sound conversions, and acoustic recording-to-sound conversions. These methods were different pathways, but they produced the same physiological effect when testing similar material. This was the moment the idea expanded. The effect was not tied to a single medium. It appeared whenever the underlying pattern was preserved.

The Compact Disc Experiment

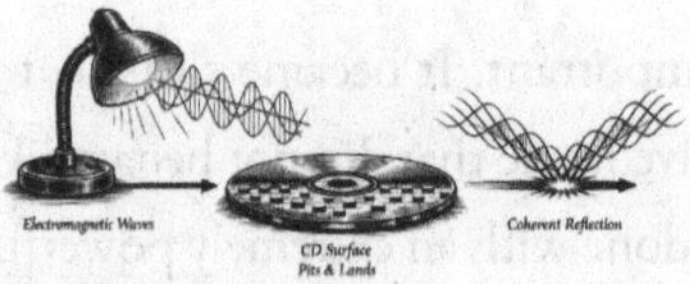

Another important clue came from an experiment using a compact disc. I placed a lamp near a disc that contained thousands of small photographs of an herb. Each photograph was converted into binary data as pits and lands burned into the surface. Those pits and lands contained the secondary information that described the pixel arrangement in the photographs, and that pixel arrangement came from the photograph that corresponded to the molecular magnetic field of the herb. When the lamp light reflected from that patterned surface, the previously chaotic light became coherent. I confirmed the match through the physiological effect they both produced. A more detailed explanation appears later in the book.

The compact disc experiment was my attempt to create an automated frequency generator that did not need me to interpret and compose the frequencies. The goal was to remove myself from the process so that anyone could set it up and still produce the desired physiological effect. It would simply emit coherent light into the space. Essentially, that was as far as I could go. There was no way to determine the frequencies without me getting involved in the process.

Coherent Reflection in Natural Light

Another early observation came from noticing how sunlight behaved when it reflected off natural surfaces. I did not understand the mechanism at the time, but certain areas in nature produced a distinct physiological effect when I was physically present. The effect was not tied to brightness or appearance. It depended on how the molecular structure of the terrain shaped the reflected sunlight, even though I did not yet have the language for it.

A separate observation appeared indoors. When light from a lamp or electromagnetic field from a cellphone playing music reflected from a test object, the reflected field produced the same physiological effect. The Third Rail continuity test (my sensitivity) confirmed that the frequency pattern was present in the reflected field as an analog of the molecular magnetic field.

These observations showed that coherent reflection could appear in natural sunlight and in artificially generated electromagnetic fields when they interacted with a molecular surface, as explained in more detail in Optical Chapter 3 and validated through the testing procedures in Chapter 8.

The 2018 Breakthrough

That changed in 2018, when my friend Mark Ohlson introduced me to a program called AudioPaint that could convert a photograph directly into an audio file. When the audio file was played through the electrical audio coil, it produced the same physiological effect of the item as what I created in a tone generator or from direct handling of the material. This confirmed that the pattern of the molecular structure survived the transformations across these media and opened the path to creating more than 1200 analogs of herbs and medications. I later switched to a more powerful program called Audacity, which became the stable platform for generating these audio files.

Why I Continued My Amateur Research

During this same period, I worked with individuals with developmental disabilities in support roles at Seattle Children's Hospital. I did not directly observe treatment, but I was regularly in areas where children with serious medical conditions were receiving care. The experience showed how limited the available options can be for many families, even with advanced medical science. That reality became one of the primary reasons I continued pursuing the avocational investigation described in this book alongside my regular job. These early observations formed the first trail of pattern-based clues. They showed that the same pattern was appearing across electromagnetic analogs and sound-wave analogs, two different media showing the same pattern-based frequency detail. They also showed that the physiological effect was a reliable indicator of when that pattern was present.

Chapter 2: The Logic of Coherence

Why molecular structure produces a reproducible frequency pattern

In this book the term top-down states that coherence begins outside the medium and outside creation as the intelligibility order that gives rise to the pattern. Coherence appears first as testable in physical form at the covalent molecular level as the frequency pattern of the molecular magnetic field. The medium does not create this pattern. It can only preserve it or degrade it. Top-down describes the origin of the coherent frequency pattern and the way the pattern appears in physical form. It does not describe direction or layers or any form of material hierarchy.

Modern bottom-up science often interprets molecular structure as the statistical result of many small interactions that are considered random. Bottom-up science explains molecular behavior using internal probabilistic interactions, and it does not recognize or model anything that persists when the pattern appears in another form, such as an optical, acoustic, or electromagnetic pathway.

The Logic of Coherence begins with the opposite assumption. It holds that coherence is force presented as a frequency pattern, that this pattern remains the same when represented in different media, and that it is not random. Coherence originates in the atomic and molecular domain. No other medium can generate coherence. It is the underlying intelligible pattern. This persistence across media is not produced by bottom-up interactions and shows intelligibility. As pixels or binary data, coherence appears as secondary pattern detail, and as an electromagnetic field and as sound waves it remains force presented as a frequency pattern in the natural world.

Sub-frequencies

Coherence is presented as a frequency pattern of the molecular magnetic field, and the strongest points of that pattern appear as peak frequencies in the electromagnetic analog. Sub-frequencies are smaller components that lie below the peak frequencies and make up the overall frequency pattern of the same coherence. They are sub-analogs of the molecular magnetic field and reflect finer detail within the covalent frequency pattern. I do not analyze the sub-frequencies because I have no method to isolate or evaluate them independently. Peak frequencies are sufficient for identifying the overall coherent frequency pattern that has a physiological effect and is testable.

Coherence Originates Only in Molecular Structure

Every molecular structure has a specific arrangement of atoms, and that structure produces the molecular magnetic field. The bonds, electron distribution, and bond geometry form a stable pattern presented as the molecular magnetic field, which is force as a frequency pattern. The field is not classical magnetism as it is viewed today. It is a frequency that shows the same pattern as the covalent pattern.

In this book coherence frequency patterning is observed indirectly through frequency analogs of the molecular structure. It is the frequency pattern generated by the molecular magnetic field, and this pattern is tied directly to the geometry of the molecular material, the covalent structure. Historical technologies such as the phonograph and telephone were not designed to work with coherence. They were intended only to convey information such as words, music, and images. Whenever the source had covalent molecular structure, the analog presented coherence, but no one recognized it at the time. Symbols do not contain coherence frequency patterning because a symbol is an image, not a molecular structure.

Reversibility in Mapping and Designing Molecular Magnetic Fields

When the molecular magnetic field of a covalent structure appears as an electromagnetic field analog, the analog preserves the coherence peak frequency pattern of the original field. This preservation allows researchers to examine the peak frequencies directly, because the pattern remains tied to the geometry of the covalent molecular structure that produced it. The testing chapter discussed later in this book provides a template. The mapping between covalent structure, the molecular magnetic field, and the peak frequency pattern can therefore be used in both directions. The forward direction begins with the covalent structure and proceeds through the molecular magnetic field to the peak frequency pattern that appears in the analog. The reverse direction begins with the preserved peak frequency pattern and proceeds back toward the covalent structure that corresponds to it.

Because this mapping is reversible, researchers can analyze the preserved peak frequencies to reconstruct the covalent molecular structure. They can compare related structures by examining differences in their peak patterns. They can also design new covalent structures by adjusting the peak frequency pattern and then analyzing the effect of those modifications. The purpose of these modifications is to produce a covalent structure whose molecular magnetic field is more effective for the outcome they want.

Forms of the Coherent Molecular Magnetic Field

The coherent molecular magnetic field can take on many secondary encoded forms, including binary patterns, optical patterns, and artificial-intelligence--generated image patterns, while not losing coherence. Electromagnetic fields and sound waves are the primary force-based pathways that present coherence as analogs of the molecular magnetic field. Secondary encoded media can be translated back into these force-based pathways by a physical interface using an electrical audio coil.

How Coherence Survives Loss of Detail

Coherence can be presented as a stable molecular magnetic field that can transfer across different media, even when the representation of that pattern is degraded. The specific elements of how degradation occurs are not known, though it is believed to appear in mirror images, computer screen images, and artificial-intelligence-generated images when these are processed through a photograph-to-sound program for analysis. Because coherence does not depend on the completeness of the representation, a degraded form can present a detectable coherence pattern.

Coherent Electromagnetic Field Interaction with Molecular Structures

Electromagnetic fields reflect from some materials and pass through others, and in many situations, they do both. These pathways are ordinary physics and not the focus of this book. What is new is the observation, confirmed through Third Rail continuity testing, that when an electromagnetic field interacts with a molecular structure through reflection, transmission, or a combination of both, the interacting field becomes coherent.

This effect was seen in tests using light from a lamp and electromagnetic fields from music on a cell phone interacting with the test items. The same effect explains the coherence that appears when sunlight reflects from natural surfaces. In each case, the reflected electromagnetic field becomes coherent because of its interaction with the molecular structure.

The Electrical Audio Coil as a Dual Medium Translator of Coherence

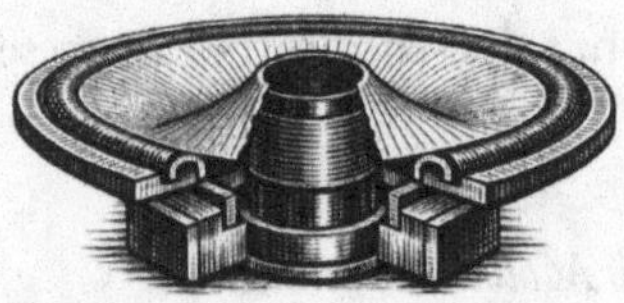

The electrical audio coil is a simple solenoid-based device, part of the same lineage as telegraph sounders and doorbells, later refined into the modern speaker. It is a coil of wire that is loosely suspended around a bar-magnet assembly, with a small air gap between the bar and the coil where the magnetic field is concentrated. The coil is driven by an alternating current (AC) audio signal generated by an audio file. The voltage rises and falls in a patterned waveform that corresponds to the timing and frequencies encoded in the sound file. As the AC signal flows through the coil, it interacts with the magnetic field and produces an oscillating force that moves the coil up and down in step with that waveform. When a diaphragm is attached, this motion produces a sound wave analog of the same pattern. When no diaphragm is attached, the coil still produces an electromagnetic analog of that pattern.

The electrical audio coil preserves the pattern of the coherent molecular magnetic field when that pattern is carried as an electrical signal. When the field pattern is present in the electrical signal, the coil produces two physical analogs of the same pattern at the same time. One analog is electromagnetic field frequencies, and the other is sound wave frequencies when a diaphragm is attached. These are not different signals. They have the same coherence appearing in two different physical media. This is why the coil belongs in the Logic of Coherence. It demonstrates that the same field pattern can appear in different media without being altered.

Artificial Intelligence Images and Coherence

Artificial intelligence (AI) image generation begins with patterns learned from large sets of photographs. These photographs contain pixel patterns that present coherence in optical form if the training photographs were of actual material. When the AI model generates a new image, it reconstructs a pixel pattern based on these learned relationships. Even though the image reconstruction is synthetic, it presents a degraded but identifiable form of coherence that can be processed as a molecular magnetic field analog in a photograph-to-sound program. This has been demonstrated in the testing chapter.

AI images can be used in a similar manner to photographs of real molecular structures. The specific way coherence is preserved through artificial intelligence image generation is not fully understood, but the resulting images when processed as molecular magnetic field analogs create similar physiological effects.

Chapter 3 The Optical Pathway: The Layout of Coherence

How reflected light presents the molecular magnetic frequency pattern

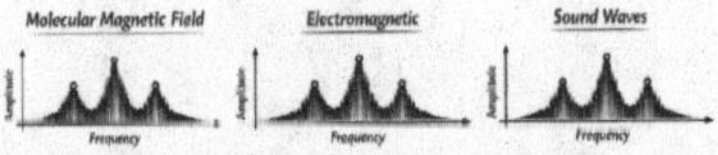

The previous chapters established that a molecule's geometry produces a covalent molecular magnetic field that is preserved as an electromagnetic analog and an acoustic analog when played through an electrical audio coil. They are two physical presentations of the same molecular magnetic field frequency pattern because the electrical audio coil produces both at the same time. This chapter explains how coherent reflected light presents a coherent frequency pattern shaped by the covalent geometry of the material, and how a matrix of pixel values preserves that coherent frequency pattern.

A photograph does not contain force. It holds secondary information in the form of a pixel arrangement that preserves the field pattern of the molecular magnetic field. This secondary information is similar to the early phonograph mechanical groove to sound representation discussed at the beginning of the book.

Why Reflected Light becomes Coherent

When light reflects from a material, the photons interact with the electrons that form the covalent molecular structure. Those electrons are arranged in a fixed geometric pattern, and that pattern does not change when photons arrive. The energy of visible light is too low to eject electrons, too low to break bonds, and too low to alter the covalent molecular geometry. Instead, the photons interact with the electron distribution, become coherent as an analog of the molecular magnetic field of the material, and leave again. Because the electrons remain within their covalent structure, the reflected light carries a stable frequency pattern shaped by that covalent molecular design. The light becomes coherent to the geometry it encountered. This dynamic follows the same logic as the CD experiment, except that instead of a secondary encoded pattern created by pits and lands, the actual molecular structure is the pattern.

The Optical Pathway

When coherent reflected light is captured by a camera, the pattern becomes a pixel numerical pattern. The numerical pattern can be translated into binary code, and binary can be translated into different physical media. Pixels, pits and lands, magnetic domains, and audio samples are all physical translations of the same field pattern of the molecular magnetic field as long as the pattern is preserved in the new format. These different media are secondary coherence pathways.

The photograph-to-sound program performs one of these translations. It transforms the numerical pattern of the photograph into a frequency pattern that forms an audio analog. When played through an electrical audio coil both the electromagnetic field analog and sound wave analog produce coherence as force when the pattern is preserved. A spectrum analyzer displays the strongest (peak) frequencies in that audio file so they can be extracted.

Digital Sampling and Pixel Structure

Before the program converts a photograph into a sound file, the reader needs to understand one idea. A digital photograph is a grid of numbers called pixels, and those numbers in the pixel structure preserve the pixel-level relationships created by the material's covalent geometry. The program reads the numerical pixel values, including the color channels, and converts those values into sound according to a fixed rule without analyzing objects, shapes, or scenes. When the numerical structure preserves the pattern shaped by the material's covalent geometry, that pattern can be transformed into a coherence frequency. This is why the photograph-to-sound program works, because it treats the photograph as a table of values whose internal relationships correspond to the pattern produced by the molecular magnetic field.

How a Real Photograph Appears as Numbers

These numbers are an example of the pixel values the camera program imports as raw data from the photograph.

[10, 12, 14]

[11, 15, 13]

[12, 14, 16]

This matrix is not a picture. It is a table of values. The important feature is the pattern formed by the relationships between the numbers. These relationships, which include gradients, boundaries, and variations, show how the frequency pattern appears in numerical form. Color values are not required for preserving the coherence pattern, but when a photograph contains color, those values are part of the pixel data and flow through the numerical pipeline the same way as any other editable pixel values.

Color as a Coherent Electromagnetic Frequency

Color is electromagnetic frequency, and any stable color such as red from paint, fabric, a digital display, or a sunset presents a frequency pattern that can be coherent. Color is produced by molecular or atomic structure through reflection, emission, or transmission, and the structure determines the electromagnetic frequencies that appear as color. The pixel values preserve the numerical relationships created by the coherent frequency pattern of the molecular magnetic field. Black-and-white photographs preserve the same coherence pattern as color photographs of the same material because coherence depends on those numerical relationships shaped by the covalent geometry, not on color. The material that produces the color contributes to the optical information, and the degree to which this contribution affects the coherence pattern is unknown.

This allows the program to convert those numerical relationships into a frequency analog. The physiological effect comes from the coherent electromagnetic frequency presented by the optical information of a material and not from the material in a symbolic pattern. Color cannot enter the acoustic pathway because microphones detect only sound waves, but color can enter the electromagnetic tuning pathway when the optical information is converted into an electromagnetic analog.

A Close-Up Example Showing a Simple Structural Pattern

Pixel matrices contain visible structures. The image below shows a small block of nine pixels, and it demonstrates a simple structural pattern created by one value that differs from the others. This single example makes the structural pattern easy to see because the difference is clear.

[10, 10,10]

[10, 30,10]

[10,10, 10]

The center value differs from the surrounding values. Each number is one pixel, and this 3×3 block is a small section of a photograph. This is a structural pattern. The program does not interpret this as brightness or contrast. It interprets it as a set of numerical relationships. The pattern is created by the difference between the center pixel and the eight surrounding pixels, and this difference is what the program reads.

These numerical relationships preserve the frequency pattern that corresponds to the coherence describing the molecular magnetic field, and this is why the program can use them during the conversion process. A photograph preserves this pattern because the pixel values preserve the pattern produced by the molecular magnetic field created by the material's covalent geometry.

Pixel Arrangement and its Frequency Analog

A pixel value is a numerical sample of reflected light, and coherent reflected light preserves the field pattern produced by the molecular structure. A frequency value is a numerical sample of sound. Both are numerical forms of the reflected light. When a pixel arrangement preserves this field pattern, the program can translate that numerical structure into a frequency structure. The comparison below does not assign specific acoustic frequencies to the pixel values. Pixel values are samples of coherent reflected light, not sound, so they do not contain acoustic frequency peaks. The program uses the numerical differences between the values to create a mapped frequency pattern, and the mapped frequencies depend entirely on the program's conversion scale. The pattern is preserved because the relationships between the pixel values remain the same, not because the numbers contain predetermined frequency peaks.

Pixel arrangement: [10, 30, 10] This row contains a repeating pattern: low, high, low.

Program mapping: 10 becomes a lower frequency, 30 becomes a higher frequency

Resulting sound wave structure: low → high → low

Mapped frequency structure (conceptual): [f_1, f_2]

The exact numbers depend on the program's mapping, but the pattern is preserved.

The structural pattern in the pixel values becomes the structural pattern in the mapped frequencies. The program does not create the pattern. It transfers it. Pixel structure becomes frequency structure because the program converts one into the other.

The Compact Disc Experiment as a Subset of the Optical Pathway

The compact disc experiment discussed in Chapter 1 demonstrates that coherence can be preserved and presented in a different physical form while still following the same logic as the optical pathway. The disc does not originate coherence. It preserves the frequency pattern already present in the photographs encoded into it, and this preservation allows the optical pathway to continue from the disc surface without requiring a photograph-to-sound program.

A compact disc surface contains pits and lands that encode the pixel arrangement as binary values. The pits and lands form a line of values, for example: [1, 0, 0, 1, 1, 0, 1]. This binary row is a physical representation of the same content found in a pixel row. It presents the same content in binary form, and the binary stream is equivalent to a digital image array once it is decoded.

In the compact disc pathway, the photographs have already been encoded as a binary stream that preserves the pixel arrangement. The coherent reflected light from the disc presents the preserved pattern directly. This makes the compact disc pathway a subset of the optical pathway because it does not require conversion to an audio analog through the photograph-to-sound program.

The reflected light from the lamp does not require an electrical audio coil interface. The electromagnetic field produced by the reflected light has the same physiological effect as the photographs encoded into the compact disc. At present, this can only be confirmed through the Third Rail continuity test.

Chapter 4: The Acoustic Pathway

How micro-vibrations of a material, appearing as near-surface sound waves, mirror molecular magnetic fields

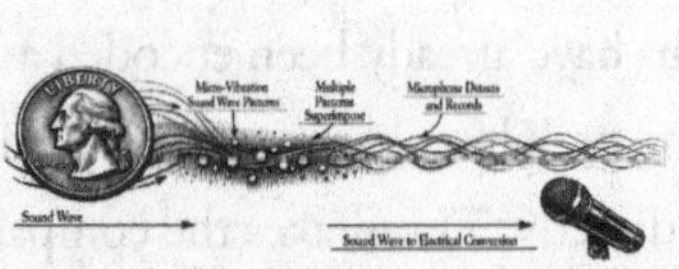

In this book coherence that is testable begins at the atomic and molecular level. This points toward the quantum domain, which is not testable and is covered in Chapter 9. The covalent geometry of atoms and molecules determines the coherence frequency pattern that appears in every pathway described in this work. This pattern originates in the molecular geometry and remains the same when it appears in different physical forms. This top-down frequency pattern is the foundation of coherence. The molecular magnetic field pattern and the micro-vibration sound wave pattern both originate from the same geometry. The acoustic pathway is one of the three pathways through which coherence becomes observable.

Molecular Structure and its Two Pathways

Because atoms are always in motion, the molecular geometry produces a continuous micro-vibration sound wave pattern. This vibration is extremely small, but it is stable and tied directly to its geometry. The same molecular geometry also produces a molecular magnetic field pattern. These two patterns originate from the same geometry and remain stable as long as the geometry exists. Both patterns can be replicated as sound analogs and electromagnetic analogs without altering the coherent frequency pattern.

Micro-Vibration of Materials

The covalent geometry of a material vibrates at micro-scale and reflects the same coherence frequency pattern that appears in the material's molecular magnetic field. That is, they both present the same coherent frequency pattern in two related media. Because both micro-vibration and the molecular magnetic field arise from the same covalent geometry, a material's vibration can be recorded as patterned ambient sound waves in the air boundary near the material. These sound waves propagate outward into the surrounding air. When recorded and analyzed in a spectrum analyzer, peak frequencies can be extracted that correspond to the coherent frequency pattern of the material's magnetic field. Larger materials may radiate more cumulative micro-vibration sound waves due to increased surface area, and microphone proximity can influence how much of this vibration is captured.

Microphones and Micro-vibrations

Microphones are designed to detect extremely small changes in sound wave pressure. The microphone records the ambient noise that contains the sound wave pattern produced by the material. This is possible because sound waves are in superposition, which means multiple sound wave frequencies are present at the same time but do not blend. The microphone transforms the sound waves into a coherence-encoded electrical signal. When the recording is processed in a photograph-to-sound program, the audio is converted into a sound file that can be analyzed in a spectrum analyzer, allowing peak frequencies to be extracted. These peak frequencies correspond to the molecular magnetic field frequency pattern of the material. This is the acoustic pathway in its simplest form. The electrical signal contains the frequency pattern of the material.

Reversibility of the Electrical Audio Coil

The electrical audio coil is central to why acoustic recordings preserve coherence. In a speaker, an electrical signal drives the coil, the coil moves the diaphragm, and the diaphragm produces sound. In a microphone, sound waves move the diaphragm, the diaphragm moves the coil, and the coil generates an electromagnetic signal. Because the coil is fully reversible, it preserves the frequency pattern in both directions.

Multiple Materials in the Same Recording

Each material in an area produces its own unique sound wave pattern. Because sound waves behave linearly under normal conditions, they do not overwrite one another or merge into a single waveform. They superimpose. Superposition shows that each sound wave pattern remains intact even when many are present at the same time. This is the same physical principle that allows a microphone to record an entire orchestra without losing the distinct sound of each instrument. Noise-cancelling systems do not interfere with this process. They cancel only large, predictable, repetitive components of ambient noise. They do not cancel sound wave patterns because these patterns are extremely small, non-periodic, and not predictable. The acoustic pathway therefore remains valid even in environments where noise-cancelling technology is present.

Why the Acoustic Pathway Matters

The acoustic pathway matters because it shows that micro-vibration patterns from a material can be recorded as sound and converted into a coherence-encoded electrical signal. That signal can then be translated into an electromagnetic or audio analog that produces the same physiological effect as the material and has been confirmed in testing discussed in Chapter 8. This demonstrates that the coherent frequency pattern of the molecular magnetic field can appear through micro-vibration as sound waves and can appear in electromagnetic and acoustic forms. The acoustic pathway therefore provides direct evidence that a frequency pattern can remain stable across media while still producing a physiological effect.

Coherence of sound waves that describe the magnetic field of a material, as presented in this chapter, is not part of any established scientific framework. It is a novel phenomenon with profound implications, and its consistency has been confirmed through testing discussed later in this book.

Chapter 5 Why the Human Body Responds to Coherent Frequency Patterns

How coherent frequency patterns create a physiological response

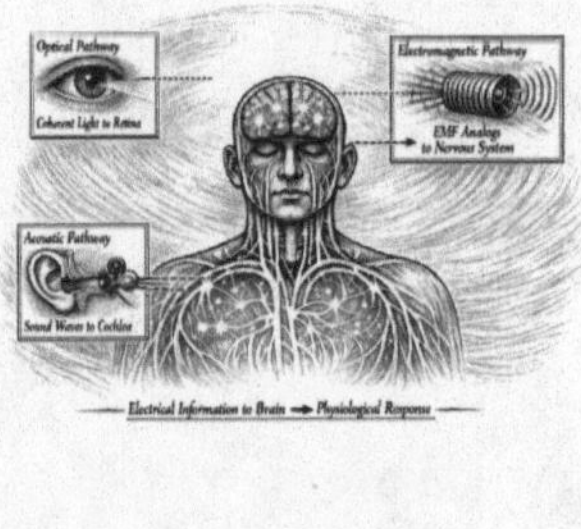

The human body responds to coherent frequency analogs because it is an electrically active system built on processes that change state when electrical conditions shift. Coherent electromagnetic field analogs preserve the same frequency pattern as the molecular magnetic field of the original substance. When these analogs interact with the body, the electrical systems that regulate physiological function shift in ways that reflect the applied electromagnetic field analog. This chapter explains why this interaction is possible.

The Human Body Is an Electrical System

Human physiology is fundamentally electrical. Many essential processes operate through electrical changes that trigger or regulate biological activity. Cells maintain voltage differences across their membranes. Ion channels open or close when electrical thresholds are crossed. Neurons fire when depolarization reaches a specific voltage. Heart rhythms, brain rhythms, and sensory pathways operate through electrical timing. Even protein structures can shift when local electrical conditions change.

Electrical and magnetic fields are inseparable. Every electrical process in the body is accompanied by a magnetic component, even though scientific descriptions often focus on the electrical aspect. These systems do not require strong forces. They respond to small, structured electrical inputs when those inputs interact with their thresholds. Because the body is electrically active, it is capable of responding to electromagnetic patterns, not just chemical substances.

A New Way to Understand the Human Electromagnetic Field

The human body emits several electromagnetic fields, including low-frequency electrical fields, magnetic fields generated by electrical activity, and thermal infrared radiation. As discussed in previous chapters, a molecular structure creates a covalent electromagnetic field that can be preserved in another medium as an electromagnetic analog presented through an electrical audio coil.

Each emission type presents the full individualized molecular magnetic field patterns of the human body in totality. This full pattern is coherence of the molecular magnetic field appearing consistently across different emission types. A useful way to understand this is to imagine a hologram. When a hologram is broken into pieces, each fragment still contains the entire image. In the same way, each emission type, whether low frequency electrical fields or thermal infrared emission, presents the full pattern associated with the body's individualized molecular structures.

In another related category, sensations such as pain, touch, and certain emotional responses arise from electrical patterns generated by the emotional state, and electrical patterns always include electromagnetic components.

Pathways Into the Human Body

An electromagnetic field analog is the coherent frequency pattern of a targeted molecular magnetic field. It influences electrical pathways in the body that the brain processes as something internally occurring such as ingesting an herbal supplement. The brain does not know that the stimulus comes from outside in this example. The electrical audio coil is the central interface in this work. It receives a coherence-encoded electrical signal that already preserves the pattern of the targeted molecular magnetic field. The coil presents that signal in two forms. It produces an electromagnetic field analog and a sound wave analog, and both expressions present the same pattern that the brain interprets as an internal process.

Human Eye (Optical Pathway): Reflected coherent light from molecular structures enters the eye, and the retina produces electrical impulses that preserve the pattern. Optical pathways remain valid in this system because coherent light preserves the frequency pattern.

Human Ear (Acoustic Pathway): Mechanical sound waves are converted into electrical impulses in the inner ear and are sent to the brain, entering the same physiological domain as the optical pathway.

Human Nervous System (Electromagnetic Pathway): The nervous system is electrically active throughout the body, and its electrical activity provides a pathway for electromagnetic coherent frequency patterns to interact with and influence receptive structures in the physiological domain.

The Nervous System as the Body's Unified Electrical Network

The body's electrical systems do not operate as isolated parts. Every electrical process, whether it begins in a cell membrane, a sensory receptor, or a muscle fiber, feeds into a single integrated network governed by the nervous system. At the center of this network is the brain, the organ responsible for receiving, interpreting, and coordinating electrical activity throughout the body.

The brain is the body's primary electrical processor. It regulates heart rhythm, breathing patterns, hormonal release, digestion, immune responses, movement, and every form of perception. All of these functions depend on electrical signals that the brain continuously receives and adjusts. Nothing in the body operates without this electrical oversight.

When a signal enters the body from the outside world, it is converted into electrical information before it reaches the brain. Light entering the eye becomes electrical impulses in the retina. Sound entering the ear becomes electrical impulses in the cochlea. Subtle electromagnetic fields can be converted into electrical information by tissues that are already electrically active. Once converted, all of these signals travel along the same nerve pathways that lead to the brain.

Because the brain integrates every electrical signal it receives, it does not distinguish between the sources of those signals. It processes electrical information based on structure, timing, and pattern, not on whether the signal originated from a molecular structure, a sensory input, or an external coherent electromagnetic field analog. The nervous system treats all electrical information as part of the same continuous flow.

The electromagnetic field analog is the coherent frequency pattern of the targeted molecular structure. When it interacts with the body, it moves into the same electrical pathways that the brain uses to regulate physiological function. The brain receives this information and incorporates it into its ongoing electrical activity.

Sensory Responses to Coherent Frequency Analogs

The three pathways into the body discussed above create an avenue for sensory replication using coherent frequency analogs. Sensations such as heat, cold, smell, taste, and emotional responses can be reproduced when the frequency analog correctly matches the electrical signal associated with that sensation that travels to the brain for processing. For example, an emotion creates a specific electromagnetic frequency that can be identified and replicated through a photograph or acoustic recording.

Historical Snapshot of the Natural Environment

Another way to consider the natural environment is as a complex frequency pattern formed by the molecular magnetic fields of materials in that area at that time and place. This includes the electromagnetic field generated by the electrical pattern of an emotional state and other sensations of an individual in the photograph. In this regard the photograph can be considered a historical record of molecular magnetic field patterns that can be extracted, analyzed and classified.

Future Research Considerations

Although coherent electromagnetic field analogs preserve the frequency pattern of the original molecular structure, the detailed internal mechanism by which the body interprets this pattern is not yet known. Validating that these coherent frequencies have physiological counterparts is one of the goals of this book. Future researchers will need to determine how these analogs are best designed for maximum physiological effect. This includes identifying the most effective duration of exposure, the frequency of application, and the conditions under which the structured electromagnetic pattern produces the clearest physiological response. These questions extend beyond the scope of the current work and will require systematic investigation.

Chapter 6 Imprinting

How receptive materials acquire a modified molecular magnetic frequency pattern

This is the first chapter where coherence acts directly on matter, and where the coherent frequency pattern of a molecular material modifies the molecular magnetic field of another. This is unusual because a molecular magnetic field is normally fixed by covalent molecular structure. Modifying it without chemistry is not part of classical physics or classical magnetization.

Imprinting as a Form of Magnetization

Imprinting is a form of magnetization. It is a pathway for applying a coherent frequency pattern into the molecular magnetic field of a target material. Classical magnetization uses a strong, chaotic electromagnetic force to align magnetic domains. Imprinting uses an electromagnetic analog that is being applied to the receptive material. The applied force is not chaotic. It is coherent. When this frequency pattern interacts with a receptive material, the material's molecular magnetic field is modified and now has field elements of both the applied molecular structure and the receptive molecular structure.

My imprinting work used an electrical audio coil playing an electromagnetic analog of an herb or supplement. The Third Rail continuity test allowed me to detect when a material had taken on a coherence pattern. The 36-sector radial testing in Chapter 8 confirms continuity between imprinted material using a photograph-to-sound program as a photograph and also a recording of the target material. The process follows the top-down order that coherence is presented as a frequency pattern, and imprinting is the application of that pattern as force into another material. Because imprinting produces a long-lived modification of the receptive material, the effect points to a change occurring within the molecular structure itself at an atomic level, possibly involving a shift in spin dynamics.

Why Modifying a Molecular Magnetic Field Is Novel

Classical physics does not describe a pathway for modifying the molecular magnetic field of a receptive material with the physiological effect presented by an applied electromagnetic analog of an herb or supplement without changing the material itself. Classical magnetization aligns magnetic domains but does not modify the molecular magnetic field defined by the covalent molecular pattern. Imprinting is different. It shows that a coherent frequency pattern can modify the pattern of a material's molecular magnetic field. This is the first demonstration in this book of top-down force acting directly on matter. It is unusual because the modification is a form of magnetization that is not the result of chemistry. The coherent frequency pattern becomes part of the material's molecular magnetic field. This is why imprinting is significant. It shows that coherence is not symbolic. It is physically consequential.

How Imprinting may Modify a Molecular Structure

Imprinting applies a coherent frequency pattern into the molecular magnetic field of a receptive material. Because the molecular magnetic field is produced by the covalent molecular geometry, any change in the field indicates a proportional change in the molecular structure that generates it. This relationship is defined as reversibility. Classical physics has no pathway for producing this kind of modification without altering the material itself. Imprinting is different. If the applied coherent frequency pattern modifies the molecular magnetic field, then the molecular structure may have been modified in a corresponding way. This may represent a top-down effect on matter, and it is potentially consequential. It shows that the relationship between the covalent molecular geometry and the molecular magnetic field may be reversible.

Low power imprinting produces a limited modification of the molecular magnetic field of a receptive material. High power imprinting may produce a deeper modification because a stronger applied field can exert greater influence on the receptive material. A high-powered electromagnetic analog may produce a more pronounced modification of the receptive material's molecular magnetic field. If a greater modification of the molecular magnetic field occurs, then by reversibility it may indicate a corresponding change in the receptive material's molecular geometry. When this occurs, the receptive material may take on a more stable coherence pattern, the modified field may persist longer, and the physiological effect may be more detectable.

Why Imprinting is Currently Subtle

My work at YouTube Helpful Waves for almost twenty years has shown that electromagnetic analogs of herbs and supplements often produce clear physiological effects for many listeners. Imprinting produces effects that are much more subtle. The reason is the molecular magnetic field of the target material and items discussed in the previous section. The receptive material retains its original field, and the applied coherent frequency pattern is added to it, producing a hybrid field. The hybrid field is never as pristine as the electromagnetic analog itself. Future research will refine this process, but imprinting was not my primary focus and remains an area of further discovery.

Receptive Materials

A receptive material is a material whose molecular magnetic field can be modified by a coherent frequency pattern. The idea of a receptive material is not fully defined, but it can be described as a molecular structure with a non-dominant molecular magnetic field. A non-dominant field does not overwhelm the applied coherent frequency pattern. My imprinting work focused on materials that were susceptible to this form of magnetization. I primarily used aluminum blanks and silicone wristbands. Plastics were found to have impurities that produced unwanted physiological effects. Though my testing in this area was limited, the size of the imprinted material may also be a factor in the imprinting receptivity.

Retention and Re-imprinting

One of the most interesting discoveries is that materials can retain their modified molecular magnetic field for years. Once a material has been imprinted, it does not return to its pre-imprint state. The imprint becomes part of the material's molecular magnetic field. Similar to classical magnetism, an imprinted item can be re-imprinted with different electromagnetic analogs. The new imprint overwrites the previous imprinting, and the material's molecular magnetic field reflects the new coherent frequency pattern. Another observation from my work is that the resulting molecular magnetic field often appears stronger than the material's original pre-imprint state.

The Question of Structured Fixed Magnet Creation

Imprinting led me to a larger question about magnetization itself. Classical magnets are created using strong, chaotic electromagnetic force. The force aligns magnetic domains, but it does not preserve any covalent frequency information. The result is a magnet with a magnetic field that reflects the alignment process, not coherence.

The open question is whether a magnet could be created using an electromagnetic analog instead of a chaotic force. If a magnet were manufactured using a frequency analog as the applied force, the resulting magnetic field might contain coherence. This would be a new category of magnetic material. If this were possible, it would be a significant development and would mean that coherence could be embedded directly into a magnetic material during its creation. This remains an open question and a direction for future research.

Chapter 7 Periodic Table of Peak Frequencies

Mapping molecular structure through stable peak frequency patterns

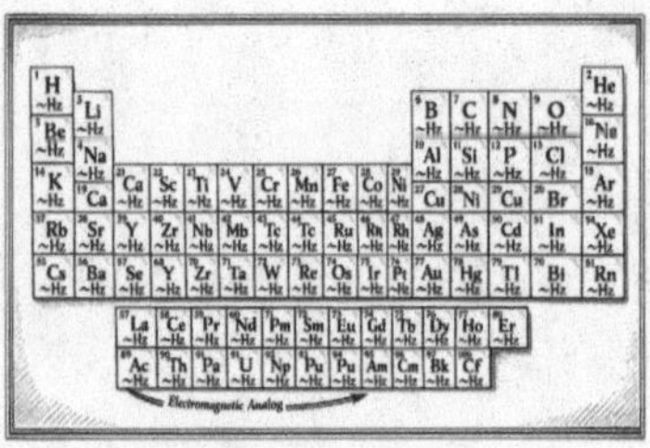

Coherence and top-down architecture described in this book make it possible to represent each element in the periodic table in terms of its peak frequencies. The periodic table of elements is normally used to organize the elements by atomic number, electron arrangement, and chemical behavior. This traditional layout helps scientists understand how elements bond, how they form molecules, and how they behave in reactions. By adding peak frequencies as a complementary layer, each element gains an additional dimension based on the frequency pattern produced by its covalent atomic structure.

When an element is associated with its peak frequencies, researchers can study how these patterns relate to covalent molecular structure. This provides a direct way to identify how each atomic component contributes to the coherent frequency pattern of the molecular structure. The table of peak frequencies does not replace the periodic table. It builds upon it in a way that supports further scientific investigation.

When elements are combined into molecular structures, their individual peak frequencies are not mixed or blended but remain part of a coherent covalent frequency pattern of the molecular structure. When this happens, individual element peak frequencies do not lose their distinct pattern. Instead, the coherent frequency pattern of a molecular structure reflects the arrangement of these atomic components rather than a replacement of them. This is similar to tuning a radio to locate one station. Each station on the frequency band keeps its own signal while contributing to the overall composite pattern.

Peak Frequencies and Sub-frequencies

In this chapter, we focus on peak frequencies of electromagnetic analogs of molecular magnetic fields and how they correlate to the elements. In a more refined sense, this is only the surface description of the electromagnetic analog. Beneath the peak frequencies, as shown on a frequency spectrum analyzer, are the sub-frequencies that form the overall pattern. These sub-frequencies are sub-analogs of the same molecular magnetic field and represent the atomic and molecular contributions that together form the total covalent molecular magnetic field.

Radial and Spherical Audio Analog Models

The spatial layout of peak frequencies can also be represented through radial and spherical audio analog models. The 36-sector radial model is a visual layout of peak frequencies arranged in sectors to make the coherence pattern easier to visually observe for material comparisons in testing. The spherical model extends this idea into three dimensions and represents the same coherent frequency pattern as a surface distribution around a sphere. Both models show how the peak frequencies and sub frequencies form a stable pattern that remains consistent when converted into an electromagnetic analog or a sound analog. These models do not change the underlying frequency data. They provide a visual and spatial way to observe how coherence organizes itself when represented across different geometries. They are representations of their respective molecular magnetic field counterparts as coherent frequency analogs. Further information on how radial and spherical analog models are used in testing is presented in Chapter 8.

Chapter 8 Testing for Coherence

Using the 36-sector radial method to visually compare frequency patterns

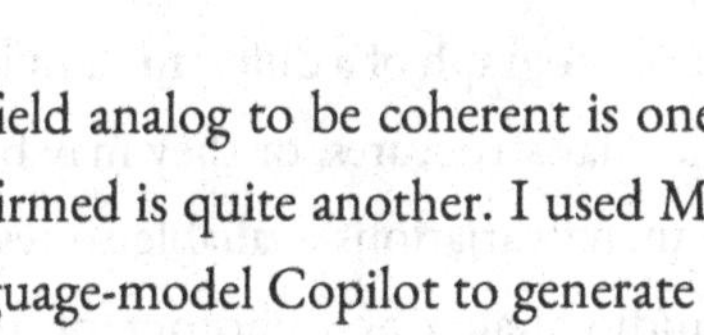

This chapter presents the testing method that provides data that may validate my hypothesis when evaluated by researchers. The results are preliminary and very promising. Claiming an electromagnetic field analog to be coherent is one thing and showing how coherence is confirmed is quite another. I used Microsoft artificial intelligence (AI) large-language-model Copilot to generate the testing output under strict controls and oversight.

My goal in this chapter is to show the testing method clearly and precisely. Appendix A will provide more details for the exact testing setups. The key to all of this is electromagnetism in the form of an electromagnetic field analog. The stable platform for this work is the electrical audio coil and a photograph-to--sound program. I currently use Audacity. With these tools, the testing method becomes reproducible and structured.

The electrical audio coil is the component that makes the entire testing method possible. In a speaker, a coherence-encoded electrical signal drives the coil, the coil moves the diaphragm, and the diaphragm produces sound. In a microphone, sound waves move the diaphragm, the diaphragm moves the coil, and the coil generates an electromagnetic signal. Because the coil is fully reversible, it preserves the frequency pattern in both directions. The peak frequencies extracted from the resulting audio file can be examined directly with a spectrum analyzer and form the basis of the testing template used to evaluate coherence. Although molecular magnetic fields can be measured in principle, this work cannot measure them directly.

The question addressed in this chapter is whether two different media preserve the same frequency pattern. The testing template is designed to determine whether a photograph of a molecular structure and an audio recording of coherent ambient noise near the material present the same coherent frequency pattern in the form of peak frequencies. If my testing template is validated by future researchers, it would indicate that both media preserve the same pattern. The protocol and testing template described below will help researchers evaluate whether cross-medium coherence is real. This chapter focuses exclusively on peak frequencies because they are the only measurable coherent frequency pattern available for testing, apart from my subjective Third Rail continuity check.

One can compare one photograph to another photograph of a different sample type. They may not be the same class of molecular structures, or they may be the exact same targeted material. There are many variations available to test for coherence. One can also compare the audio analog of a photograph to coherent ambient noise recorded near the material in various combinations. These comparisons determine whether two materials share the same molecular pattern in the form of peak frequencies.

Because the electrical audio coil converts the frequency pattern directly into electromagnetic force, a coil-only device can produce a coherent electromagnetic field analog without producing sound. Removing the diaphragm does not change the frequency pattern associated with the coil, as it only removes the acoustic output. In electromagnetic form, the peak frequency analog appears as a purely electromagnetic signal. This shows that the electromagnetic field preserves the same frequency pattern even when the acoustic pathway is removed.

Photograph-to-Sound Testing Platform

The photograph-to-sound program is the foundation of this testing method. Audacity is the program I use. Other programs will no doubt work if they can convert photographs into sound and display a spectrum. Audacity accepts both photographs and acoustic recordings. It converts each one into a sound file that can be analyzed. The identified peak frequencies were also entered into a tone generator for a second Third Rail continuity test for confirmation.

Everything in the Testing Method Depends on the Peak Frequencies

Everything in the testing method depends on the peak frequencies extracted from the audio file. These peak frequencies are the only measurable pattern available for comparison. They are the part of the electromagnetic field analog and sound wave analog that can be examined directly. Every comparison and every evaluation of continuity or divergence use these peak frequencies as the basis for analysis. Without peak frequencies, there is no way to determine whether two molecular structures during testing are similar or dissimilar.

During testing I use specific terms for the final analysis of the material comparisons. Continuity is the observable match between two peak frequency patterns. Continuity appears when the peak-frequency pattern is preserved with only local differences. It ties directly to coherence as an observable feature of the frequency patterns. That is, it is coherence presented as numerical data in the test. Divergence is the opposite of continuity and is the observable loss of the material peak frequency profile. Divergence appears when extra frequencies or missing frequencies occur in the testing with a marked change in the overall frequency pattern and will be explained in more detail later in this chapter.

Radial and Spherical Representation of the Electromagnetic Field Audio Analog

The 36-sector radial system used in this work is derived from dividing the 360-degree equator of a sphere into 36 equal parts, but the testing did not use radii, angular coordinates, or any spherical calculations. All peak frequencies are maximum value points and therefore belong on the outer radius of the sphere at the equator. Even if angular coordinates were computed, every peak would still lie on the same equatorial ring because the underlying data is one-dimensional. For this reason, the equator was treated as a straight frequency line and unwrapped into 36 equal frequency sectors for testing. Sub-frequencies are the only values that could move north or south in a future spherical model. The radial and spherical diagrams in this chapter are idealized visual layouts of the same 36-sector structure and not computed coordinate systems. The testing itself used only the straight-line sector assignments.

As discussed above, the peak frequencies extracted from the spectrum analyzer were the key to testing for coherence. The issue was how to find the pattern that demonstrated continuity or divergence when comparing peak frequencies of materials either as photographs or sound recordings. It proved to be a daunting endeavor with many false starts. As this work was done primarily by matching frequencies and looking for patterns, testing took a long time to find the correct setup. Initially it was hoped that listing frequencies side-by-side as close neighbors would offer the desired information within frequency sectors. The offsets were arbitrarily derived. This basic methodology proved to be unworkable after repeated renditions. The displayed data did not show coherence across media consistently.

This led to a new presentation for the peak frequencies in a radial format using sectors. 144 sectors were initially used, then 72 and finally 36, with a frequency spectrum divided between 0 and 10,000 hertz. Even with those comparisons, the testing failed to produce consistent results. There were too many peak frequencies to evaluate using Hann window at FFT log 2048. At that point, it seemed unlikely that consistent results could be achieved with the tools available.

The key that eventually proved successful was selecting the log number that was not too condensed and not too expanded. The new insight was combining previous peak frequencies into a more condensed yet informative set that still displayed enough peak frequencies suitable for comparison. This issue is analogous to combining the peak frequencies in an ordered way that showed the top of the hill and not the tiers to the summit. The frequency spectrum was also increased from 10,000 to 25,000 hertz. FFT log 256 proved to be the sweet spot for testing using a reduced radial sector count of 36.

The 36-sector radial format was used to organize the peak frequencies extracted from the spectrum analyzer. In the larger spherical layout, of which the radial sectors are a subsection, the term "radial" refers only to how the 36 sectors appear when wrapped around the sphere. The testing itself uses the flat 36-sector layout derived from the equator. Each primary peak frequency is placed on the equatorial ring of a sphere as a reference marker for that material's coherence pattern.

Spherical Model

In this model, the peak frequencies occupy the outer radius at the equator, and the associated sub-frequencies are shown as offsets extending along the north–south axis of the corresponding longitude. These offsets are mirrored as positive and negative values because the model shows polarity mirroring the polarity of the covalent molecular magnetic field of the material. When all the sub-frequencies and peak frequencies are placed into this spherical layout, the result is a full three-dimensional representation with fixed coordinates for every frequency in the model. This produces a clear sector pattern that can be compared directly to the patterns of other materials in future research. Materials that share the same underlying frequency pattern produce peak frequencies that fall into the same sector regions.

36-Sector Radial Testing of Audio Analogs

The 36-sector radial test consistently identified continuity or divergence between materials. Most material pairs produced distinct sector patterns that were either aligned or separated in ways that were easy to interpret. Testing showed that a photograph of a material and an acoustic recording of the same material produced peak frequency sets that remained stable within a bounded range. This stability confirmed that both pathways preserved the same coherent frequency pattern.

When two materials are compared, the goal is to see whether their peak frequency patterns show continuity or divergence. The sector layout makes this comparison visual and more straightforward, though the patterns require discernment when analyzing. Each material produces a pattern of occupied and empty sectors, and the way these patterns line up shows whether the two materials can be classified as having the same underlying molecular structure or not.

The first step is simply to look at the overall sector patterns. If the two materials fill the same sectors, continuity is likely. If they occupy different sectors, divergence is clear. Most comparisons fall cleanly into one of these two categories.

However, some materials produced close matches, and these required a second level of analysis. In these cases, more detailed features were considered such as:

Missed boxes: sectors where one material should have a frequency but does not

Asymmetric multi-frequency sectors: places where one material has several frequencies and the other has none

Single-material sectors: sectors occupied by only one material

Frequency-count differences: how many frequencies appear in each sector for each material

large frequency differences in the sectors

Other fine detail considerations:

One to one match: one frequency in each material's box. This is continuity.

One to many match: one frequency in one material, multiple in the other. This is continuity because both materials occupy the same sector. However, if one-to-many patterns appear repeatedly across multiple sectors, this may become a divergence signal depending on other testing factors.

Many to many match: both materials show multiple frequencies in the same sector. This is continuity because both materials occupy the same sector.

Empty: one material has frequencies, the other material's box is empty. This is divergence.

Staggered empty: both materials have frequencies, but in different sectors. This is also divergence.

These finer details helped confirm whether a close match is truly continuous or only initially appears similar. The final judgment of continuity or divergence is based on the overall pattern along with the secondary criteria used when two materials look nearly identical. This ensures that every comparison is evaluated consistently and that borderline cases are resolved with clear, objective rules.

Detailed Sector Analysis Criteria

As discussed above, some testing comparisons required a more detailed level of analysis. Because the evaluation is visually determined, there is a degree of subjectivity in how these finer frequency patterns are recognized. The testing was not performed without knowing the materials, so the extent of this subjectivity is unknown and likely impacts testing where materials show close frequency pattern profiles.

One testing indicator is differences in sector occupancy. Two materials may appear to match, yet upon closer inspection show partial occupancy, uneven fill patterns, or sectors that look aligned but differ in frequency count.

A second indicator is stair stepping in the frequency progression. In a clean continuity pattern, the peak frequencies increase in a smooth, consistent progression through the 36-sector radial pattern. When the progression jumps unevenly or forms step-like intervals, the sequence of peak frequencies is less consistent between the two materials.

A third indicator is reversion in the frequency progression, where a later peak frequency falls into a lower sector than the one before it.

A fourth indicator is multi-frequency imbalance within shared sectors. Two materials may occupy the same sector, but one may show a single peak frequency while the other shows several, or both may show multiple peak frequencies that do not match.

Materials Test

Material	Material
14 kt gold ring photograph	14 kt gold ring mirror reflection
14 kt gold ring recording	14 kt gold pendant photograph
Amber photograph	Amber recording
Amethyst photograph	Amethyst recording
Jalapeño pepper photograph	Jalapeño pepper recording
Lemon photograph	Lemon mirror reflection
Lemon recording	Fine silver coin photograph
Fine silver coin mirror reflection	Fine silver Standing Liberty coin photograph
AI generated image of .999 silver bar	

The 14 kt gold ring photograph, 14 kt gold pendant photograph, and 14 kt gold ring recording all showed continuity with one another. All 14 kt gold items diverged from amber, amethyst, jalapeño, lemon, and silver.

Amber

The amber photograph and amber recording showed continuity with each other and divergence with all other materials.

Amethyst

The amethyst photograph and amethyst recording showed continuity with each other and divergence with all other materials.

Jalapeño pepper

The jalapeño photograph and jalapeño recording showed continuity with each other and divergence with all other materials.

Lemon

The lemon photograph, lemon mirror reflection photograph, and lemon recording all showed continuity with one another and divergence with all other materials. The lemon photograph and the fine silver photograph were analyzed in more detail and confirmed to be divergent. This was performed using the secondary more refined analysis discussed above.

Fine Silver Coin (.999)

The fine silver coin photograph, Standing Liberty silver coin photograph, fine silver coin mirror-reflection, and fine silver coin recording all showed continuity with one another and divergent with all other materials. The fine silver photograph and the lemon photograph were analyzed in more detail and confirmed to be divergent.

AI-generated .999 Silver Bar

The AI-generated silver bar image showed degraded continuity with the fine silver coin photograph and divergence with all other non-silver materials.

Introduction to the Test Tables

The following tables present three partial examples from the 36-sector radial testing method used to evaluate continuity and divergence between materials. This gives the reader an idea of what the test was looking for. For comparison testing, all 36 sectors are evaluated. Each full sector table lists all 36 sectors and shows where the peak frequencies lie in the sectors with the materials being compared. The markers used in evaluation are consistently used across all sectors.

Continuity Table — Fine Silver Coin Photograph vs Fine Silver U.S. Walking Liberty Coin

Sector	Fine Silver Coin Photograph	Fine Silver US Walking Liberty Coin
7	4571	4433
8	5064	5002
9	5760	5801
11	7322	7597
20	13387	13730
21	14136	14539

The Fine Silver Coin and Fine Silver U.S. Walking Liberty Coin photographs show a strong continuity pattern across the 36-sector test. Out of 36 possible sectors, 29 sectors are filled by both materials, with only 3 partially filled sectors and 4 empty sectors. The frequencies rise smoothly through the sectors with no marked increase in frequency differential between adjacent frequency sectors large enough to break the expected incremental progression. The six sectors shown here illustrate the shared structure clearly, with both materials occupying the same sectors across a wide range of the radial model.

Continuity Table Amethyst Photograph vs Amethyst Recording

Sector	Amethyst Photograph	Amethyst Recording
14	9116	9099
15	9882	9824
16	10878	10782
19	13078	13031
20	13589	13693
24	16622	16633

These six sectors are clean, tight, and representative of the overall pattern. The amethyst photograph and amethyst recording show a strong continuity pattern across the 36-sector radial test. Out of 36 possible sectors, 32 sectors are filled, and most of these are occupied by both materials. The six sectors shown here illustrate this continuity clearly. In each case, the photograph and recording frequencies fall into the same sector with only small local differences in value. This high level of sector occupancy and closely aligned frequencies demonstrates the stable continuity expected when two coherence pathways preserve the same underlying frequency pattern.

Divergence Table — 14 kt Gold Ring Photograph vs Amber Photograph

Sector	14 kt Gold Ring Photograph	Amber Photograph
3	1582, 1998	—
6	4097	—
16	10507	—
27	—	18289
29	—	19530, 20002
30	—	20427

The sectors shown below were selected because they demonstrate clear single--material occupancy. In sectors 3, 6, and 16, only the 14 kt gold ring photograph contains peak frequencies, while the Amber photograph has none in those sectors. In contrast, sectors 27, 29, and 30 contain frequencies only from the Amber photograph, with no corresponding values from the 14 kt gold ring. These examples illustrate how the sector table identifies divergence by showing sectors in which only one material is present.

Continuity and Divergence

The Fine Silver comparison demonstrates how two objects made of the same material can produce a continuous and stable sector alignment. The Amethyst comparison extends this further, showing that continuity can appear even when a photograph and an acoustic recording of the same material are tested. In both cases, the sector patterns remain consistent, stable, and repeatable across the full set of 36 sectors. <u>Finding frequency coherence between a photograph and an acoustic recording of the same material is a new discovery.</u>

Two mirror tests were also analyzed using the testing protocol and they both demonstrated continuity, confirming that coherence survives optical reflection. The fine silver coin mirror test was based on the image being displayed on a computer monitor, while a real lemon was used in its mirror test. The difference in continuity of the reflection source was found to be significant. Coherence was degraded with the Fine Silver Coin test and remained stronger with the lemon.

The divergence examples in this chapter show the opposite pattern. When two materials do not share sector structure, the 36-sector radial test makes the mismatch visible. In the 14 kt gold ring photograph versus Amber photograph comparison, sectors that contain frequencies from only one material, along with staggered empty patterns where each material occupies different sectors, indicate that the two materials are not the same. These patterns show that the test can separate materials cleanly, and that divergence is detectable even when the frequencies appear close in value.

Material Imprinting Testing

Testing for continuity and divergence in imprinted material was successful. Imprinting using a low-powered coherent electromagnetic field as the applied force on the receptive material was expected to show degraded coherence. The reason is the now-imprinted material has elements of both the applying molecular magnetic field as an electromagnetic field analog and the receptive material's molecular magnetic field. This modified field is tied to what I believe to be a change in the spin dynamics of the molecular structure. Something professional researchers can sort out.

As mentioned in the previous imprinting chapter, imprinted material keeps its new modified field for years and has a physiological effect similar to the imprinting analog. The Third Rail continuity test has confirmed this by both hands-on and testing with audio analogs of the material.

In the testing procedure a silicone wristband was imprinted with an electromagnetic field audio analog of the Fine Silver Coin photograph. A second non-imprinted silicone wristband was used for testing control. The Fine Silver Coin imprinted wristband was also recorded by a microphone. Photographs of the imprinted and non-imprinted wristbands and the imprinted wristband's acoustic recording were processed with Audacity and the peak frequencies were extracted. The analysis of the testing generally required a deep level of visual analysis. However, one test comparison produced a level of continuity frequency alignment that exceeded expectations.

Test Results Imprinted Silicone Wristband

Sector	Fine Silver Coin Photograph	Fine Silver Coin Wristband Imprinting
10	6509, 6917	6523, 6877
11	7322	7634
12	7795	8197
13	8408	8734
14	9090, 9671	9098, 9720
16	10866	10599
17	11793	11316
18	12373	12056, 12479
19	12674	12888
20	13387	13633
21	14136	13982
22	14807	14794
23	15298, 15828	15319
26	17384, 17953	17528
28	18861, 19435	18883
29	19837	19809
30	20743	20392
31	21144	20940
32	22153	21708
34	23522	23442

The test comparison results shown above produced an unexpectedly strong result. Even though the Fine Silver Coin Photograph and the Fine Silver Coin Wristband Imprinting are not the same material, the majority of their sectors are populated, and they move through the sector sequence in almost perfect parallel, showing strong continuity. The average frequency difference per shared sector is 269 Hz and the median is 289.5 Hz, which means the material sector frequencies remain very similar across both lists. This level of alignment is unusual for different materials and demonstrates that silicone is a highly receptive medium and an excellent material for imprinting.

Test Results Imprinted vs. Non-Imprinted Control Wristband

Sector	Imprinted Fine Silver Coin Wristband	Non-imprinted Wristband (Control)
1	—	587
3	1659	—
6	—	3949
9	5760, 6100	—
16	10866	—
24	—	16056
34	23522	—

Sector	Imprinted Fine Silver Wristband Recording)	Non-imprinted Wristband (Control)
1	—	587
5	2949	—
7	4266, 4608	4811
14	9175, 9866	—
26	—	18110
29	19835	—

In the full 36 sector table comparing the Imprinted Fine Silver Coin Wristband to the non-imprinted wristband, the pattern initially appeared to show continuity because both materials occupied many of the same sectors. That superficial overlap made the two patterns look closer than they actually were. However, when the full sector pattern was analyzed, the divergence became clear. The imprinted wristband filled 29 sectors, the non-imprinted wristband filled 26 sectors, yet only 21 sectors were shared, leaving 13 sectors where only one material appeared. Also, the imprinted wristband showed more multi frequency sectors and some sectors showed wide frequency differentials. When taken as a whole, the sector pattern met the definition of divergence.

Similar to the previous comparison, the test between the Imprinted Fine Silver Coin Wristband Recording and the non-imprinted wristband initially appeared to show continuity because both materials occupied many of the same sectors. That superficial overlap made the two patterns look closer than they actually were. However, when the full sector pattern was visually analyzed, the divergence became clear. The imprinted wristband filled 29 sectors, the non-imprinted wristband filled 26 sectors, yet only 21 sectors were shared, leaving 14 sectors where only one material appeared. The imprinted wristband also showed more multi frequency sectors and some sectors showed wide frequency differentials. When taken as a whole, the sector pattern met the definition of divergence.

Continuity Confirmed by Parallel Divergence

When the two Fine Silver Coin imprinting comparisons (photograph and recording) to the non-imprinted wristband are reviewed side by side, the data shows a repeated sector pattern that adds significant weight to their earlier testing. The testing results demonstrated continuity between the imprinted Fine Silver Coin wristband and the Fine Silver Coin wristband recording. This current testing method using a non-imprinted wristband as a control shows that same continuity relationship from another angle.

Test	Imprinted Wristband	Non-imprinted Wristband (control)	Shared Sectors	Unshared Sectors
Recording vs Non-imprinted	29	26	21	14
Imprinted vs Non-imprinted	29	26	21	14

In the testing table above, the imprinted wristband and imprinted wristband recording diverged from the non-imprinted wristband in many of the same sectors. Both tests produced nearly identical sector counts, with the imprinted wristband filling 29 sectors and the control filling 26 sectors, and both produced 21 shared sectors with 14 unshared sectors. Both also showed multi--frequency behavior in similar regions, including sectors 7, 8, 14, 20, 22, 26, and 28. These repeated numerical patterns show that the imprinted Fine Silver Coin photograph and Fine Silver Coin recording break away from the control wristband in parallel ways, reinforcing the previous testing results of their continuity.

A Newly Discovered Imprinting Effect

One of the imprinting tests had a significant outcome that affects how continuity and divergence should be interpreted. Photographs of a Fine Silver Coin and an imprinted Fine Silver Coin silicone wristband produced peak frequencies that matched through the sectors, even though both materials are dissimilar. This is remarkable, as the imprinted material showed no outward visual signs of the internal molecular magnetic field modification. This result shows that imprinting can, in some situations, produce outcomes outside the expected behavior of the 36-sector radial method.

Some Thoughts on the Spherical Format and Coherence

The spherical layout may reveal consistent angular relationships among the frequency peaks, patterns that become more visible in three dimensions. These relationships are scientifically important because the same frequencies shown in this model produce a measurable Third Rail physiological effect identical to the effect of the molecular structure itself. This demonstrates that the coherence pattern is not arbitrary but reflects the same organization presented by the covalent molecular magnetic field. The 36-sector radial spherical format maps all peak frequencies and sub-frequencies into a fixed global arrangement, revealing frequency relationships that other methods may not show.

The proposed model reflects the organization of the covalent molecular magnetic field produced by the molecular structure itself. This book has already established these relationships, and the hypothesis defines the coherence pathway clearly. The electromagnetic frequency analog describes the same covalent molecular magnetic field that arises from the molecular structure, and that structure originates at the quantum level, which will be examined in the next chapter. The eigenstates in the quantum domain occupy defined regions, and the spherical model provides a fixed coordinate system for displaying the coherence pattern that may originate from those regions. This proposed pathway for coherence extends from the subatomic to the atomic and molecular level.

Chapter 9 Quantum Coherence and the Deeper Order

How molecular coherence appears in the quantum domain

The following is my hypothesis that I do not have the ability to test. I am simply attempting to show the reader a possible pathway of coherence in this domain.

Coherence is a top-down pattern that precedes the quantum domain. It is the pattern outside of creation that the quantum domain follows. Quantum mechanics does not create coherence. It may provide a physical domain where coherence appears through nonmixing eigenstates, which are quantum states with one definite value. Coherence appears at the quantum level because the quantum engine follows the pattern. When this definite value arises from a definite energy of excitation, one possible avenue for this energy may be the gluon-based interaction. This definite energy corresponds to a single mathematical frequency that may be mirrored as a peak frequency in the 36-sector radial presentation of the pattern.

The Order of Creation

Modern science teaches that the world is built from the bottom-up. The quantum world is said to generate atoms. Atoms generate molecules. Molecules generate the physical world. In the top-down view used in this book, quantum mechanics is not the source of coherence. It is the mechanism that brings coherence into physical form. Coherence exists first as a coherent pattern outside creation. It is the first structure and is the template that defines what can exist inside creation.

The quantum domain is the engine that produces physical form. The pattern precedes the particles. The coherent pattern may provide the bridge between the quantum and the natural world and shows how the deeper order predicted by Bohm, Penrose and others becomes visible. This is the foundation of this chapter.

This means that the separate and nonblended nature of quantum mechanics is not an accident. It is not a property that emerges from particles. It is the physical presentation of a deeper pattern that is already separate and nonblended. The quantum domain may present this pattern through eigenstates. Each eigenstate may correspond to coherence as a peak frequency presented in the physical world. Each eigen region mirrors the logic of the peak-frequency sectors by sharing the same non-mixing rule, while the sectors themselves remain a representational coordinate system used for displaying the coherence pattern.

The quantum world is not building coherence. It is following an intelligible pattern. The pattern is primary. The mechanism follows. The quantum engine is secondary. This is why the coherence system described in this book and the quantum system show the same behavior. Both are separate. Both are nonmixing. Both are complete. Both are stable. They show these properties because they share the same origin.

The simplest way to understand this relationship is to see that coherence is the blueprint. Quantum mechanics is the machinery that follows the blueprint creating physical matter. The blueprint does not originate from the machinery. The machinery follows the blueprint.

Bohm and the Implicate Order

David Bohm argued that the visible world unfolds from a deeper order. He called this the implicate order. Bohm believed that the apparent randomness of quantum mechanics hides a guiding structure. He insisted that the world is not built from isolated particles but from a continuous underlying pattern that shapes what we see. Bohm never identified the pattern of this deeper order. He only predicted that it must exist. Bohm's concept of the pilot wave is central to this deeper order and may be directly related to the coherent frequency patterns described in this book and explained in more detail at the end of the chapter.

Penrose and the Objective Collapse Framework

Roger Penrose argued that quantum state collapse is objective and physical rather than a product of measurement. He believed that collapse is driven by deeper structure and that quantum mechanics is incomplete without a unifying order that connects it with general relativity. In the type of deeper structure Penrose predicted but did not specify, the coherent frequency pattern of the molecular structure described in this book may reach into the quantum world as a mirror of that relationship. The pattern sets the allowed states from the top down, and this is the kind of deeper structure Penrose argued physics is missing. Penrose anticipated a deeper order, but he did not provide the data to validate it. This book provides a testing template that identifies the pattern, shows how it determines allowed states, and may offer the missing explanation Penrose argued must exist.

Eigenstates in Quantum Mechanics

An eigenstate is a definite quantum condition. It holds one value and one condition, remaining stable and not combining with other states. Each eigenstate belongs to a complete set that spans all possible configurations. These properties make eigenstates the fundamental units of quantum behavior. A quantum region can contain several allowed eigenstates, but only one becomes the outcome. That outcome eigenstate remains separate, nonmixing, and unchanged across boundaries.

The eigenstate rule is important because it may be the link that mirrors peak frequency behavior in the 36-sector radial model representing the molecular magnetic field of the material. These frequencies cannot blend or overlap since combining them is physically impossible.

Singleton

The singleton is the primary peak in a 36-sector radial system used for testing. It is the one peak in each sector that anchors that sector's contribution to the overall coherence frequency pattern and remains stable across materials and media. The singleton observation shows the coherence frequency match because coherence testing can allow several possible peaks with one primary singleton peak, and quantum mechanics can allow several possible eigenstates with one becoming the outcome. A sector in the 36-sector radial coherence pattern is an allowed region where peaks can occur, and a quantum eigen region is an allowed region where eigenstates can occur. The important rule is that boundaries do not mix so frequency peaks do not cross sectors and eigenstates do not cross eigen regions.

Gluon Confinement and the Non-Mixing Rule

The behavior of the gluon offers one possible avenue for understanding how a definite quantum excitation can remain stable. In quantum mechanics, the gluon maintains the binding of quarks by preventing separation and preventing mixing across boundaries. This behavior may reflect a nonmixing rule at the subatomic level. The gluon does not allow partial combinations or blended states. The pattern it maintains remains whole.

This possible parallel matters because the coherence frequency pattern shows the same rule. The molecular magnetic field holds a stable frequency pattern that does not mix with other patterns. The gluon is not the source of this coherence pattern. It is simply one candidate for how a definite quantum excitation can acquire a definite energy that remains stable. This stability is consistent with the nonmixing behavior that appears in the coherence pattern at the macroscopic level.

Hamilton's Rule and the Stability of the Frequency Pattern

The Hamiltonian defines the energy structure of a quantum system. When a system occupies a definite energy state, the Hamiltonian produces a single, definite frequency. This is the Hamilton effect in which a stable energy value corresponds to a stable frequency that does not mix with other frequencies. The result is a peak frequency that remains intact across time and across presentations.

This rule may be relevant for understanding how the coherence pattern appears in the quantum domain. The gluon provides the definite energy of excitation. The Hamilton effect converts that definite energy into a definite frequency. Because the frequency is single and nonmixing, it produces a peak that remains stable. This peak may be the presentation of the coherence pattern within the quantum domain.

The importance of the Hamilton effect in this chapter is that it explains why the coherence pattern remains stable when it is rendered into physical form. The quantum domain does not generate the pattern. It presents the pattern by following the Hamilton rule in which definite energy produces definite frequency, and definite frequency produces a stable peak. This is the mathematical basis for the nonmixing behavior that allows the same coherent pattern to appear in both the quantum presentation, where the pattern is shown through eigenstates, and the molecular presentation, where the pattern is shown through peak frequencies.

36-Sector Radial Model and Spherical Model Mirroring the Quantum System

The 36-sector radial model, as a subsystem of the proposed spherical model, mirrors the quantum system because both arise from the same deeper order outside of creation referred to as top- down. Quantum mechanics presents the coherent pattern into physical form and brings the atomic and molecular structures into appearance. These structures have covalent bonds, and the covalent molecular pattern of the atoms in the molecule produces a molecular magnetic field with a specific frequency pattern. This frequency pattern is a presentation of coherence, and coherence does not change.

At the quantum level, the pattern is presented through eigenstates. At the atomic and molecular level, the same pattern is presented as peak frequencies. Both levels show allowed regions that mirror each other. Both levels have nonmixing boundaries. This is not merely analogy. It may be the same coherent pattern appearing at the quantum level. Bohm's concept of the pilot wave may be the quantum-level presentation of this deeper order, and the relationship may be explored in either direction because the pattern does not change across levels. Coherence is presented at both levels and is the template. Quantum mechanics follows this template and produces the physical structures that show it. Because both the pilot wave and the 36-sector system can be given mathematical expression, the 36-sector radial model used in testing and the spherical model as its fullest expression may give researchers a glimpse into the formation of the pilot wave.

Chapter 10 Coherence in Spacetime

How gravitational coherence appears in planetary spacetime

The following is my hypothesis, which I do not have the ability to test. My aim is simply to show the reader a possible way coherence could operate in spacetime.

Gravity links all celestial bodies through continuous mutual influence. Each body generates a coherent macro-vibration pattern arising from the combined effect of its total molecular magnetic fields. Because gravitational influence is shared, these patterns are mutually present in every gravitational relationship. This creates a universe-wide network of mutual coherence, presented locally in each body as its own coherent magnetic pattern shaped by all connected bodies. This framework may be understood as a dual-mode gravitational behavior that conceptually intersects Relativity (gravity waves) and Newtonian physics.

Modern physics describes gravity through spacetime curvature, and that model reflects a revolutionary framework that continues to be confirmed by science to this day. However, the Newtonian picture of continuous mutual influence of gravity is sufficient for understanding the argument presented here. Building upon my insights on coherence, there may be another, deeper and more subtle layer to gravity that my work points to.

One of the differences between Einstein's theory of relativity and the earlier Newtonian model is how gravitational influence behaves. In relativity, changes in gravity operate as waves and travel at the speed of light. LIGO, a highly sophisticated device on Earth, detects changes in gravity as waves that indicate possible mergers of stars and black holes. Astronomers are then alerted to scan the skies for the objects. However, in the Newtonian model gravity is considered instantaneous. The Earth-Moon gravitational relationship, for example, points to a deeper consideration that may apply to our solar system and the universe. That is, there may be an underlying gravitational relationship that is real time. The following information is built on my testing that shows another pathway.

LIGO registers cataclysmic gravity waves that move at the speed of light. Other gravitational changes such as the pull from exoplanets or the tidal motion in binary stars reach Earth through their light signatures, and radial astronomy is the method that detects these changes. These effects are far too small for LIGO to register. In the dual aspect model of gravity described in this chapter a star merger gives a clear example. The gravity waves from that merger traveled through space and reached Earth today even though the event happened millions of years ago. The second aspect of my hypothesis is that the real time vibration of that system would have been present on Earth millions of years ago when the stars still existed. What we detect today is only the delayed wave from an event whose real time influence on Earth ended long ago.

The universe already updated its real-time gravitational identity millions of years ago. We are only now receiving the late-arriving report.

Planets exhibit macro-vibrations that are secondary coherent patterns described in Chapter 2. When two bodies tug on each other, the local macro-vibration reflects the coherent patterns of both planetary molecular magnetic fields. This mutual presence does not require transmission because it arises from the continuous gravitational relationship between the bodies. That is, one planet tugs and the other is immediately pulled. There is no lag time. The gravitational tug between Earth and Jupiter may be best understood as a local combined macro-vibration at each planet. What occurs at one planet occurs at the other. This vibration contains the coherent patterns of both planetary molecular magnetic fields. No transmission occurs. It is real time.

We can see the effects of this large-scale connection in many familiar places. The tides on Earth show how the Moon's presence creates a repeating pattern in our oceans and the Earth's crust. What happens to one is gravitationally happening to the other. The slight wobble of stars reveals the presence of orbiting planets. Binary stars tug each other and produce measurable fluctuations in their shared light. These are all examples of large-scale macro-vibrations presented through a gravitational connection. Because these relationships change with planetary position, alignment, and planetary orbits, the strength of gravitational coherence can increase or diminish. Gravitational changes also occur when planets themselves change, including events such as orbital shifts, collisions, or disintegration, which in the LIGO case are detected as gravitational waves.

Physicists have never connected gravitational interaction with coherent molecular magnetic field information because gravity is usually treated as smooth and structureless. However, if macro-vibrations contain secondary coherent patterns, and if gravitational relationships are continuous, then these patterns are mutually present in the shared gravitational relationship and can provide information about their respective molecular magnetic fields. I am simply implying that gravity may operate as a wave and, more subtly, as a local macro-vibration that is always in effect. This may be understood as a dual-mode gravitational behavior that conceptually intersects Relativity and Newtonian physics.

Historical technologies such as the record player demonstrate that a physical encoded structure such as an LP record can contain coherent frequency patterns carved into its grooves as secondary information. It is interpreted immediately when a needle attached to a device interacts with it, converting the secondary encoded information into coherent electromagnetism or sound waves. Micro-vibrations discussed in Chapter 4 show the same principle at the molecular scale where a microphone picks up vibrations from a material. These vibrations can later be converted into peak frequencies describing the molecular magnetic field.

Macro-vibrations extend this principle to planets, and by inference, all celestial objects, where each body has its own coherent molecular magnetic frequency pattern. Because planets continuously tug on each other, these patterns (vibrations) are mutually present in the shared gravitational relationship. My own testing suggests that such patterns can be interpreted in real time, consistent with the idea that no transmission is required, only the relationship. A vibration or tug from Jupiter is immediately a pull on Earth.

In this chapter, coherent geomagnetism plays the role of reflected coherent electromagnetic fields, allowing the same coherence principle to extend from the molecular scale to the scale of planets and the solar system. The conversion of secondary encoded information into electromagnetism, arising from the gravitational relationship between Earth and Jupiter, may resemble reflected coherent electromagnetism. Earlier chapters showed that reflected sunlight, or an electromagnetic field reflected from a material, can become coherent. In the present discussion, it is geomagnetism that becomes coherent because of the coherent macro-vibrations of Jupiter and Earth. This coherent magnetism may also have a physiological effect.

Chapter 11 Coherence, Intelligibility, and Future Possibilities

How coherence reveals intelligibility, structure, and new research pathways

Coherence began as a question about why molecular materials produced a consistent physiological effect. It became a pathway into a deeper order of reality. The earlier chapters established that coherence is presented as the frequency pattern of the molecular magnetic field presented as force in the natural world. This pattern can appear across media, and the body responds when the pattern is preserved. This final chapter turns from what coherence is to what coherence makes possible.

Because coherence is a structural property of matter, the method described in the testing chapter applies universally. Any material, environment, or object with a stable molecular magnetic field will produce its own coherent analog frequency signature. This includes terrestrial regions, manufactured products, biological materials, and even celestial bodies.

The purpose here is not to predict the future, but to outline the developments that naturally follow from a world in which coherence is real, intelligible, and transferable. These possibilities arise from the logic already established. The covalent molecular structure produces a measurable frequency pattern. That pattern can be preserved across media. Preserved patterns, either primary or secondary, can be recognized by receptive systems. The future is simply the extension of this logic into new domains.

A Library of Molecular Magnetic Field Analogs

Coherence can be described in terms of covalent peak frequencies, with the overall molecular magnetic field of the material presented as a frequency pattern. It is then possible to build a library of these patterns, stored as frequency data, spherical geometric frequency models, and audio analogs. This library would allow researchers to compare molecular structures, explore variations, and identify relationships that are not visible through traditional chemical or physical analysis. It would also allow the creation of controlled analogs for testing, refinement, and future scientific development.

A library of molecular magnetic field analogs could serve as a worldwide repository. Endangered biological materials could be preserved as coherent frequency analogs for future study. For example, a plant known for its medicinal attributes may be approaching extinction. Creating an analog with its associated frequency and coherence data preserves this information for future research and application.

Researchers can analyze these peak frequencies and use them for synthetic reconstruction of the covalent molecular structure. A coherent frequency analog therefore does more than produce a physiological response. It shows that the preserved peak frequency pattern can be used to synthetically rebuild the covalent molecular structure, even when the original material is no longer available.

Sunlight and Surfaces: Environmental Coherence

Nature has a pronounced effect on people for many reasons. The coherence of reflected light in these natural settings contributes to the physiological effect. Sunlight is a broadband electromagnetic field, and when it strikes terrain such as leaves, soil, water, or stone, it becomes a reflected coherent field shaped by the covalent molecular structure of that area. This opens the door to replicating artificial environments that are created from the real environment.

A Third Rail continuity test extended this principle indoors. Chaotic lamp light and the electromagnetic field produced by music from a cellphone both became coherent after reflecting off a molecular structure. This demonstrated that the coherence in the reflected field comes from the molecular structure. Chapter 8 testing is a controlled version of what sunlight does in nature every moment of every day, and Chapter 3 explains how reflected light becomes coherent.

Artificial Intelligence and Coherent Image Construction

Artificial intelligence (AI) systems can reconstruct images from photographs of real molecular structures. These reconstructions preserve, to a lesser degree, the molecular magnetic field's coherent frequency pattern as shown in the testing chapter and Third Rail continuity checks. As AI systems improve, they will be able to refine, adjust, and generate new analogs of these patterns with increasing precision.

AI strengthens this capability by producing reconstructions that maintain the frequency relationships needed for analysis. It allows researchers to explore variations, test adjustments, and examine how changes in the peak frequency pattern correspond to changes in the covalent structure. This opens the possibility of creating improved electromagnetic or acoustic analogs of molecular structures, including those used in medicine and assisting researchers in the study and development of medications. They provide a new way to examine frequency analogs, compare related molecular structures, and identify relationships that are not visible through traditional analysis.

Material Identification through Acoustic, Sonar, and Optical Reflections

Surface

Material identification is possible for items that are both visible and hidden when using photographs and audio recordings. Any technological device that uses light or sound records the frequency pattern shaped by the materials present. Because they preserve this material-specific frequency pattern, the coherent frequency information can be analyzed to identify structures, compare related materials, and examine features not accessible through traditional observation.

Underwater

Sound waves behave like electromagnetic fields in one important way. Both can preserve coherent frequency patterns reflected from the materials they encounter. When a sonar pulse strikes a material, or when a material is detected through passive acoustic reception, the coherent reflected wave presents the frequency pattern shaped by the molecular magnetic field structure of the surface material. Traditional sonar focuses on distance and shape, and non-coherent frequency analysis, but the reflected wave contains far more. It contains the material-specific frequency pattern that can be classified and analyzed.

Sonar may be used to detect objects by their material identity even if layered or a composite as a totality. An active sonar sweep or passive monitoring of an underwater area provides information from every object in the form of their molecular magnetic frequency analog.

Imprinting as a Form of Communication

A coherent frequency pattern can be imprinted into a material and may function as a form of communication. The imprinted frequency pattern modifies the molecular magnetic field of the material. If future researchers learn how to perform these modifications with precision, the resulting frequency patterns could be interpreted in symbolic form as alphabetical or numerical communication.

Diagnostic and System Monitoring

I often recall the old Star Trek episodes from the 1960s, where the tricorder was used to examine unfamiliar environments. In a similar way, the processes described in this book open the door to future technologies capable of analyzing molecular structures, including biological systems, through coherence-based methods. Such systems could evaluate the status and functionality of mechanical and electrical components, assess biological conditions without invasive procedures, and potentially reduce or bypass the need for animal testing. These possibilities arise from the ability to identify and compare coherence patterns across different media once scientific capabilities advance.

Chapter 12 Top-Down Creation and the Logic of Coherence

Why coherence points toward rational order rather than randomness

I have shared the history and insights gained from nearly twenty years of working with molecular magnetic field analogs. On my YouTube channel, Helpful Waves, I refer to these as designer molecular magnetic field analogs. Before turning to the broader implications, it is important to restate the central insight demonstrated through testing. Both photographs and acoustic audio can present the coherent frequency pattern of molecular structures. This finding builds upon the historical technologies discussed earlier in the book and confirms that coherence is not limited to a single medium and originates only at the atomic and molecular level. No other medium can originate coherence. This is the paradigm shift. It is a stable frequency pattern that can be preserved, transferred, and evaluated across media. The testing presented in this book shows how this works with clarity and reproducibility.

My vision for this book is to present a clear explanation of coherence and its role in producing measurable physiological effects. This explanation is grounded in the simple observation that coherence is real, that it is reproducible across media, and that the human body can interpret it with physiological clarity when the pattern is preserved.

Modern scientific reasoning often assumes that everything in creation is probabilistic and random. This assumption is presented as if it were neutral, but it requires the exclusion of logical and rational thinking to make it work. It also requires the exclusion of the obvious fact that the world is understandable. The intelligibility of creation is treated as an accident rather than a clue. As Polkinghorne observed in his work on science and theology, the deep rationality of the universe points toward a rational source rather than away from it. His point was simple. The world is not only measurable. It is meaningful.

Popular culture reinforces the bottom-up view. Books, media, and entertainment present a universe that is accidental, self-assembling, and without purpose. In this environment we are not asked to think clearly about top-down creation. We are not asked to consider that there is a creative intelligent pattern that permeates all of creation. Many scientists who hold to the bottom-up model make assumptions that exclude what is logically true. They acknowledge that the world is understandable but do not ask why it is understandable. They acknowledge that the cosmos follows patterns but do not ask why patterns exist at all.

On the other side of this question are the cultural explanations that attempt to account for human physiological sensitivity without understanding it. People experience physiological effects from material structures and assign pseudo-reasoning to the experience. They do not know that there is a logical and rational explanation for it. The purpose of this book has been to explain the logic of coherence, why it works, and how it can be validated. The human body is not guessing. It is responding to a preserved pattern. When the pattern is coherent, the response is consistent. When the pattern is broken, the response disappears.

This is why coherence matters. It is not a metaphor. It is not a symbol. It is the presentation of the frequency pattern of the molecular magnetic field. When reproduced as an electromagnetic analog or as a sound analog, it remains the same across media. The physiological effect appears only when coherence is preserved. This is not mystical. It is not speculative. It is a direct consequence of top-down order. A bottom-up world cannot explain coherence as presented in this book.

This brings us to consciousness. Science cannot explain consciousness because consciousness cannot be reduced to material processes. Attempts to do so fail because they begin with the wrong frame of reference. Consciousness is not an emergent property of matter. It is a fundamental aspect of top-down creation. Joseph Ratzinger emphasized that reason is woven into creation itself, not something that appears late in the universe. His argument was that the human mind can understand creation because creation is grounded in reason. Consciousness is not an accident. It is a gift. Roger Penrose reached this same conclusion from within physics. He argued that consciousness cannot be explained by computation or mechanism, and that something deeper, something non algorithmic and structurally prior, must be involved. His search for quantum coherence inside the brain was an attempt to locate this deeper order within physics itself. He sensed the top-down structure but could not name its source.

This is also where John Lennox's work becomes essential. Lennox has repeatedly shown that information, meaning, and rationality cannot arise from mechanism alone. Artificial Intelligence (AI) can process patterns, but it cannot originate meaning. It can generate language, but it cannot understand. It can simulate reasoning, but it cannot possess interiority. Lennox's critique of reductionism reinforces the central point of this chapter by showing that information requires a mind, and coherence requires a source.

This is why myths and misinformation about Artificial Intelligence becoming sentient spread so easily. AI is powerful, but it is not conscious. It does not possess interiority. It does not possess self-awareness. It does not possess the capacity to know. It processes patterns but does not understand them. It produces language but does not experience meaning. Consciousness cannot be understood outside a top-down frame of reference, and AI is excluded from this. It is not part of the Creator's plan. There is a plan.

The work presented in this book is one small part of that plan. It is a record of what coherence reveals about creation. It is a record of what the human body can perceive when coherence is preserved. It is a record of how top-down order makes sense of what bottom-up reasoning cannot explain. It is also a template for researchers who want to validate these findings with clarity and discipline. The logic is simple. The pathway is clear. The work is reproducible.

Creation is coherent. The human body can perceive coherence. Consciousness is real. Meaning is real. The Creator's plan is real. And the world is understandable because it was made to be understood.

My final thought is on Supreme Intelligibility, the Holy Eucharist. Many Christians struggle with the Holy Eucharist because they cannot imagine how the transformation of the bread and wine could occur without any outward physical change. Yet creation already contains examples of this in a more mundane form. At the atomic level, coherence can shift while the visible appearance remains the same. In imprinting, the pathway described in this book produces a stable change in a material's molecular magnetic field without altering how the material looks. These non-extraordinary examples, explained in Chapter 6 and Chapter 8, show that real change without outward alteration is already present in the natural world. And the change has a physiological effect. The Holy Eucharist is the supreme example of intelligibility in creation, and the work in this book has shown examples of its natural pathway into the world.

This is the conclusion of the book, but it is not the end of the work. It is the beginning of a new way of seeing what has been present all along.

Coherence Art

The world is coherent. And coherence points beyond itself to the One who spoke it into being.

References

A concise list of works that illuminate the scientific and theological context in which the ideas of coherence, and intelligibility become meaningful. These references do not define the framework presented in this book. They offer background for readers who want to explore related themes. The descriptions of these authors and their ideas are paraphrased summaries and not quotations. They are included to give context and not to provide source material for the testing model.

Bohm, David

Bohm's distinction between the implicate order and the explicate order shows how underlying quantum organization can give rise to stable, visible patterns without invoking additional forces. His view that apparent randomness may reflect deeper order offers context for readers exploring how intelligibility as coherence can emerge from ordered behavior in the quantum domain.

Wholeness and the Implicate Order. Routledge, 1980.

The Undivided Universe: An Ontological Interpretation of Quantum Theory (with Basil Hiley). Routledge, 1993.

Lennox, John

Lennox's work reinforces a top-down view of intelligibility and extends this argument into the domain of artificial intelligence. In *2084*, he shows that computation cannot generate consciousness or meaning, and that reductionist accounts fail to explain rational structure. He further argues that human rationality, moral awareness, and interiority cannot be derived from algorithmic processes, and that the distinction between symbol processing and genuine understanding exposes the limits of mechanistic accounts of mind.

God's Undertaker: Has Science Buried God? Lion Hudson, 2007.

2084: Artificial Intelligence and the Future of Humanity. Zondervan, 2020.

Merton, Thomas

Merton's Christian contemplative writings emphasize that the order of creation is grounded in God and encountered through interiority, humility, and communion. His reflections show that intelligibility is not only structural but experiential, and that meaning is received through relationship rather than constructed by the self. In *No Man Is an Island*, he highlights that the person awakens to the order of creation by participating in God's life and responding to it with love and contemplative attention.

No Man Is an Island. Harcourt, Brace, 1955.

Penrose, Roger

Penrose argues that human consciousness cannot be reduced to computation, since genuine understanding involves forms of reasoning that exceed algorithmic processes. He also proposes that quantum collapse is an objective physical event rather than something triggered by measurement, suggesting that nature contains a deeper, non-computable order. This view provides context for readers exploring how intelligibility as coherence can arise from fixed physical behavior in the quantum domain rather than randomness.

The Emperor's New Mind. Oxford University Press, 1989.

Polkinghorne, John

Polkinghorne's work articulates layered intelligibility, showing that physical, informational, and relational levels of reality coexist without contradiction. His writing provides a clear intellectual bridge between classical science and theological meaning. In his theological work, he discusses top-down and bottom-up explanations of reality, emphasizing how different levels of understanding fit together. In his physics writing, he presents Bohm's deterministic interpretation of quantum mechanics as one of several approaches to understanding quantum behavior.

Belief in God in an Age of Science. Yale University Press, 1998.

Quantum Theory: A Very Short Introduction. Oxford University Press, 2002.

Ratzinger, Joseph (Pope Benedict XVI)

Ratzinger writes that true personhood is relational to God, to others, and to the natural world. In *Introduction to Christianity*, he teaches that just as Christ the Logos is relational to the Father and the Holy Spirit in unity, the human person exists in relation with God and with his world. Creation becomes intelligible as the person shares in this divine relational grounding. This supports the manuscript's claim that intelligibility is an aspect of coherence that appears through media in a relational order that points back to the Creator.

Introduction to Christianity. Kösel-Verlag, 1968. (English edition: Ignatius Press, 1969.)

Glossary

A

Air boundary layer: The air boundary layer is the thin region of air immediately next to a material in which the air molecules become coherent by mirroring the micro-vibrations of the material's molecular magnetic field.

Acoustic Pathway: A pathway in which ambient sound waves in the air boundary layer near a material mirror the frequency pattern of the material's molecular magnetic field. When these sound waves are recorded and converted into a sound program, the resulting audio can preserve the frequency pattern associated with the material.

Ambient Noise: The natural sound environment in an area as sound waves.

Audacity (Photograph-to-Sound Program): A free downloadable program that converts photographs and recordings into sound files and includes a spectrum analyzer. It is used in this book to extract peak frequencies from optical and acoustic sources for coherence testing.

Audio Analog: A sound-based representation of a material's molecular magnetic field pattern. It can be created by converting a photograph or a recording into an audio file with a photograph-to-sound program and extracting peak frequencies with a spectrum analyzer. It can also be created with a tone generator by entering or loading the peak frequencies associated with the material.

B

Bohm, David: A theoretical physicist whose philosophical reflections pointed to a deeper relational order beneath observable events. His work suggested that the visible world is shaped by something more fundamental.

Bottom-up: The current scientific view of reality in which patterns are explained by randomness, probability, and local interactions inside a medium. This view is medium dependent, so the pattern changes when the medium changes and cannot preserve coherence across optical, acoustic, and electromagnetic forms.

C

Coherence: Coherence is presented as the frequency pattern of the molecular magnetic field produced by a material's covalent molecular structure. Coherence never changes. When it is reproduced in electromagnetic or acoustic form, it remains the same across media, although a medium can degrade it and the physiological effect degrades with it. Coherence is a representation of the molecular magnetic field and not a physical pathway. It appears in two physical domains which are quantum scale coherence at small scales and solar system and cosmos coherence at large scales. Coherence is presented in the physical domains as dual aspects of vibration and waves.

Coherent Reflected Light: The light that has interacted with a material and preserves the frequency pattern of its molecular magnetic field under specific conditions. This preserved pattern is used in testing and appears in the optical analog.

Continuity: The observable match between two peak frequency sector geometries. Continuity appears when the same material structure is preserved with only local differences. It is coherence expressed as data in the material comparison test.

Cross-media: The process of comparing peak frequencies from different media to determine whether the same coherence is preserved. It evaluates whether an optical analog, an acoustic analog, or an electromagnetic analog expresses the same peak frequency pattern as the original molecular magnetic field.

D

Divergence: The observable loss due to the change of the material peak frequency pattern. It appears when extra frequencies or missing frequencies occur in the testing with a marked change in the overall frequency pattern during sector comparison testing.

Dual-Aspect (Vibration and Wave): Coherence is presented in two physical forms. Vibration is the standing and continuous presentation of the coherent frequency pattern. Wave is the propagating and dynamic presentation of the same pattern. These two presentations arise from the same coherence and do not represent different mechanisms. The dual aspect appears in the quantum domain at small scales and in the Solar System domain at large scales.

Dual-Mode Gravitational Behavior: A conceptual interpretation in which gravity presents in two observational modes. One mode appears as a continuous and always present relational influence consistent with Newtonian gravity. The other appears as a propagating wave consistent with Relativity and gravity waves. The term describes the conceptual intersection of these two gravitational regimes.

E

Electrical Audio Coil: A wound wire device that preserves the pattern of the coherent molecular magnetic field when that pattern is presented as an electrical signal. It produces an electromagnetic analog and a sound wave analog at the same time, both expressing the same peak frequency pattern. A diaphragm is optional. The electromagnetic analog still appears when no sound is produced.

I

Imprinting Pathway: A form of magnetization. It is the pathway in which a coherent electromagnetic analog modifies the molecular magnetic field of a receptive material. The analog presents the structured frequency pattern of the original covalent molecular structure. When this structured frequency pattern interacts with a receptive material, the material's molecular magnetic field is modified. The resulting field contains elements of both the applying molecular structure and the receptive molecular structure. The material does not recreate the original molecular magnetic field. It takes on a modified field pattern produced by the interaction.

L

Lennox, John: An Oxford mathematician and internationally known speaker on the relationship between Christianity and science. His work reinforced a top-down view of intelligibility by arguing that the rational structure of the universe is not the product of randomness. His critiques of reductionism supported the understanding that meaning and relational order cannot emerge from probabilistic processes alone.

M

Media: A medium is the form through which coherence is reproduced. Media include optical forms, acoustic forms, and electromagnetic forms. The medium can degrade coherence, but it does not change it.

Merton, Thomas: A Trappist monk and contemplative writer whose reflections emphasized unity, relational meaning, and the harmony of creation. His writings highlighted that intelligibility is not only structural but experiential, and that creation is received as a coherent whole through contemplation.

Molecular Magnetic Field: The molecular magnetic field is produced by the covalent molecular structure. The arrangement of atoms, the bonds, the electron distribution, and the bond geometry form a stable pattern that is presented as a coherence frequency field. This field is not classical magnetism. It is the frequency presentation of the covalent molecular structure.

O

Optical Pathway: The pathway in which coherent reflected light preserves the covalent molecular magnetic field at the surface of a material. A camera sensor samples this field and records it as a pixel level numerical arrangement. When this pixel data is processed through a photograph-to-sound program, it produces an analog that replicates the field of the original material.

P

Peak Frequencies: The stable, reproducible frequency points extracted from the audio spectrum produced in a photograph-to-sound program. These peaks represent the molecular magnetic field as an audio or electromagnetic analog. They are the only elements of the audio file that can be examined directly. All comparison rules and all evaluations of similarity or difference use these peak frequencies as the basis for analysis. Without peak frequencies, no coherence testing is possible.

Penrose, Roger: A mathematical physicist who argued that consciousness cannot be reduced to computation or mechanism. He taught that quantum collapse is objective and independent of measurement and that a deeper order must guide physical behavior. In this book his contribution is limited to the scientific framing of non-algorithmic reasoning and objective collapse.

Physiological Sensitivity (Third Rail): The capacity to detect molecular and electromagnetic field patterns through subjective physiological response.

Polkinghorne, John: A theoretical physicist and Anglican priest who argued that the universe is structured in layers of intelligibility and that higher levels cannot be reduced to lower ones. His work emphasized that information, form, and relational structure exert real influence, and that science reveals a world grounded in order rather than randomness. In this book his contribution provides the philosophical context of intelligibility as a top-down framework for why coherence can be recognized across media.

R

Radial 36-Sector Testing Model: A geometric layout used to display the frequency values extracted from the spectrum analyzer. The 36-sector radial format is an idealized visual representation of the same straight-line frequency bins used in testing. All peak frequencies lie on the equator because the underlying data is one-dimensional, so the equator was treated as a straight frequency line and unwrapped into 36 equal bins. The model does not use radii, angular coordinates, or spherical calculations. It simply arranges the 36 sectors around a circle to show the pattern visually.

Ratzinger, Joseph: Pope Benedict XVI, a Catholic theologian, taught that creation is grounded in rational order with meaning expressed through relationships. His work emphasized that intelligibility is woven into the structure of reality and that form reveals its nature through relational meaning. In this book his contribution provides the philosophical context of Logos as a top--down framework for why coherence can be recognized across media.

Receptive Material: A material whose molecular magnetic field can be modified by a coherent frequency pattern. A receptive material has a non-dominant molecular magnetic field, allowing the applied electromagnetic analog to add to the original field and form a hybrid field.

Reversibility (Electrical Audio Coil): The property that allows the electrical audio coil to convert sound into electromagnetic signals and electromagnetic signals into sound without losing the frequency pattern.

S

Spherical Frequency Analog Model: In the spherical representation, the sub-frequency components associated with each peak extend along the north–south axis of the corresponding longitude. These offsets are mirrored as positive and negative values to complete the directional mapping. This does not imply two sets of sub frequencies. It is required for a full 3D field representation. The result is a global, directional representation of the electromagnetic analog of the molecular magnetic field.

Sub-frequencies: The smaller frequency components that lie below the peak frequencies within the overall frequency pattern of the molecular magnetic field. They are sub-frequency components of the molecular magnetic field and reflect finer detail within the covalent frequency pattern. They cannot be isolated or evaluated independently with current methods, so only peak frequencies are used in analysis.

T

Top-down: An intelligent order view of reality in which a testable pattern begins in the covalent molecular frequency pattern and not in the medium that presents it. This shows that coherence originates in the covalent molecular structure and remains the same when it appears in optical, acoustic, and electromagnetic forms. A top-down order stays stable when the medium changes, which is why analogs created from photographs, sound, or electromagnetic representations can produce similar physiological effects.

About the Author

Thomas Lee Jacobs is a Christian of the Catholic faith and an amateur scientist whose work grows out of a lifelong interest in pattern recognition of molecular magnetic fields and the information they present. He approaches scientific questions with the discipline of a careful observer and the humility of someone who understands that discovery often begins outside traditional boundaries.

Jacobs graduated from the University of Washington's NROTC program and was commissioned as a Naval Officer in the United States Navy, experiences that shaped his discipline, his sense of responsibility, and his respect for rigorous method.

His exploration of coherence did not emerge from academic laboratories, but from years of private experimentation, technical curiosity, and a willingness to follow a question wherever it led. He is the creator of the Helpful Waves YouTube channel, a long-running citizen-science project exploring designer molecular magnetic field audio analogs of herbs and supplements. Beyond this work, Jacobs contributes to the Visual Survey Group (VSG), a citizen-science project dedicated to surveying Kepler and TESS space telescope data for exoplanet and other astrophysical discoveries.

His passion for helping others led him to work with individuals with developmental disabilities. For over 24 years Tom assisted the people he served in marketing, training, and securing successful employment in their community. He lives in Missouri City, Texas, where he continues to do avocational research and write about his work.

Appendix A: Testing Template

This appendix documents the exact, arithmetic-only workflow used to convert photographs and acoustic recordings into peak frequency lists and to map those frequencies into the 36-sector radial model. It matches the final testing protocol with no deviations. Microsoft Artificial Intelligence Large-Language Model (LLM) Copilot was used to create the final testing output with strict oversight.

This testing method is intentionally simple and does not require specialized training or equipment. A photograph or short acoustic recording, Audacity, and Copilot are sufficient to produce a complete 36-sector comparison. The workflow is arithmetic-only, repeatable, and designed so any researcher can perform it without expert assistance.

An interested researcher can use Copilot to create the testing output for analysis. Use Copilot only for initial arithmetic output and confirm all conclusions through your own visual review. Copilot cannot see peak-frequency patterns during comparisons that the human mind easily sees, and this distinction is essential to understand during the initial analysis. The testing template, material peak frequencies (A and B) as tables for comparison, will have to be posted in one chat "Turn". This prevents processing errors that occur when numerical data is split across multiple Turns. Audacity or a similar program will need to be used for extracting peak frequencies.

1. Data Input Method (Photograph and Acoustic Pathways)

Photograph Import

Open Audacity.

Select File → Import → Raw Data.

Choose the photograph of the material.

Accept the default raw data import settings.

Audacity converts the photograph into a sound file.

Select the newly created track.

Note: The photograph must contain only the material with no background objects. An enlarged snip of the material was used for all tests.

Acoustic Recording Import

Open Audacity.

Select File → Import → Raw Data.

Choose the acoustic recording of the material.

Accept the default raw data import settings.

Audacity converts the recording into a sound file.

Select the newly created track.

Note: Recordings were made using an iPhone with the microphone held 1–3 inches from the material. Both pathways use the same raw data import method so the program processes the binary data consistently.

2. Spectrum Analyzer Settings

Highlight the entire sound file. Select Analyze → Plot Spectrum.

Set the following:

Window: Hann

Axis: Logarithmic

Size: 256

These settings were used for all tests and produce the peak frequency density required for sector mapping.

3. Extracting Peak Frequencies

With Hann + Log 256 selected:

View the plotted spectrum.

Identify the peak frequencies displayed.

Record all visible peak frequencies.

No smoothing, interpretation, or sub-frequency extraction is used.

The number of peaks is not fixed, but all peaks must be recorded consistently for both materials.

These peak lists form the input for the 36-sector mapping.

4. The 36-Sector Radial Model

Frequency Range

0–25 000 Hz.

Sector Count

36 sectors.

Sector Width

25 000 ÷ 36 ≈ 694.44 Hz.

Frozen Sector Formula (Required)

$k = \lfloor f / 694.44 \rfloor + 1$ Use only the rule above. No rounding. No offsets. No interpretation.

Rules:

All frequencies from both lists are mapped.

Sector numbering is 1–36.

Lower bounds are inclusive; upper bounds are exclusive.

No smoothing or invented structure.

This formula is the only allowed method for assigning peaks to sectors.

5. Sector Ranges

The 36 sector ranges define the fixed boundaries used for all tests. These ranges are not modified.

6. Sector Table Output (Print Profile)

Every sector table uses exactly three columns:

Sector <Material A name> <Material B name>

Rules:

One row per sector (1–36 fixed or only used sectors).

Column 2 lists all frequencies from Material A in that sector.

Column 3 lists all frequencies from Material B in that sector.

Use "—" if a sector has no frequencies for that material.

Material names printed exactly as provided.

No extra columns.

No A/B labels beyond the header names.

No sector math unless explicitly requested.

Arithmetic only.

No summaries.

No interpretation.

Visual Rule (Locked)

Clean, light, readable style.

Solid column lines.

Solid row lines.

No black-box code blocks.

No decorative formatting.

7. Analysis Sector (Pattern-Based)

This section evaluates the sector table using only observable arithmetic patterns. There is subjectivity in the analysis as human visual analysis is involved and no blind testing was used in the testing method.

1. Filled Sector Comparison

Identify which sectors are filled in both materials.

Identify sectors filled in only one material.

Note whether mismatches are scattered or form a pattern.

2. Progression Through Sectors

Determine whether both materials climb the sector ladder in a similar direction.

Note any runaways, collapses, or missing progression steps.

3. Stair-Stepping and Regression

Identify forward steps, backward steps, and local regressions.

Determine whether the staircase is parallel or broken.

4. Empty vs Multi-Frequency Sectors

Identify sectors with multiple peaks vs single peaks.

Identify empty sectors in one material but not the other.

Note whether these patterns are local or systemic.

Final Pattern Classification (Arithmetic-Only)

Continuity = strong overlap in filled sectors, parallel progression, smooth stair-stepping, and local-only mismatches.

Divergence = alternating filled/empty sectors, broken progression, inconsistent stair-stepping, or systemic density mismatches.

36-Sector Frequency Ranges

Sectors 1-18

Sector	Lower	Upper
1	0.00	694.44
2	694.44	1388.88
3	1388.88	2083.32
4	2083.32	2777.76
5	2777.76	3472.20
6	3472.20	4166.64
7	4166.64	4861.08
8	4861.08	5555.52
9	5555.52	6249.96
10	6249.96	6944.40
11	6944.40	7638.84
12	7638.84	8333.28
13	8333.28	9027.72
14	9027.72	9722.16
15	9722.16	10416.60
16	10416.60	11111.04
17	11111.04	11805.48
18	11805.48	12500.00

Sectors 19-36

Sector	Lower	Upper
19	12500.00	13194.48
20	13194.48	13888.92
21	13888.92	14583.36
22	14583.36	15277.80
23	15277.80	15972.24
24	15972.24	16666.68
25	16666.68	17361.12
26	17361.12	18055.56
27	18055.56	18750.00
28	18750.00	19444.44
29	19444.44	20138.88
30	20138.88	20833.32
31	20833.32	21527.76
32	21527.76	22222.20
33	22222.20	22916.64
34	22916.64	23611.08
35	23611.08	24305.52
36	24305.52	25000.00

Note: *Sector assignments are always computed by the frozen formula. Printed ranges are rounded display values only.*

Appendix B: Helpful Waves YouTube Channel

Helpful Waves is a long-running citizen science channel on YouTube that presents audio analogs of herbs, supplements, rocks and minerals, and related materials. The channel is built on the idea that a photograph of a substance contains the molecular magnetic field pattern of that material, and that this pattern can be replicated as an audio analog. When played through an electrical audio coil, the analog is presented as both an electromagnetic field and sound waves. The purpose of the channel is to observe how visitors respond to these analogs. It does not offer medical insight or advice.

Thomas Lee Jacobs is a Christian of the Catholic faith and began the project in 2007. It was revised in 2018 and now contains more than 1200 videos. These include analogs of common supplements, analogs of less familiar herbs, and instructional videos that show how to create similar analogs using freely available software. He explains that the electromagnetic field from the speaker coil and the sound waves both present the same encoded pattern. The channel encourages personal sampling of the videos and clearly states that the analogs are not medical therapy. There is no subliminal content, and listeners are encouraged to evaluate any physiological response for themselves. This approach reflects the broader idea of coherence and the role of analogs in presenting a replicated frequency pattern of molecular structures through electromagnetic field and sound waves.

Helpful Waves is one of the earliest and most extensive collections of audio analogs of herbs and supplements. It serves as a historical example of public interest in molecular magnetic field replication in an audio format and the use of photographs as source material for analog creation. It also provides context for the testing methods described in Appendix A and for the development of the ideas presented throughout this book.